Sources in the History of Mathematics and Physical Sciences 1

DIOCLES
On Burning Mirrors

The Arabic Translation
of the Lost Greek Original
Edited,
with English Translation and Commentary by

G. J. Toomer

With 37 Figures and 24 Plates

Springer-Verlag Berlin Heidelberg New York 1976

G. J. Toomer
Department of History of Mathematics, Brown University, Providence, R. I. 02912, USA

ISBN-13: 978-3-642-80983-5 e-ISBN-13: 978-3-642-80981-1
DOI: 10.1007/978-3-642-80981-1

Library of Congress Cataloging in Publication Data
Diocles (Mathematician) On burning mirrors. (Sources in the history of mathematics and physical sciences; v. 1) Translation cf Peri pyreiōn. Bibliography: p. Includes index. 1. Mathematics, Greek. 2. Conic cections - Early works to 1800. I. Toomer, G. J. II. Title. QA31.D4913 516'.23 75-38839

Sorfcover reprint of the hardcover 1st edition 1976

D. M.
Samuel Miklos Stern

Preface

This publication would not have been what it is without the help of many institutions and people, which I acknowledge most gratefully.

I thank the Central Library and Documentation Center, Iran, and its director, Mr. Iraji Afshar, for permission to publish photographs of that part of ms. 392 of the Shrine Library, Meshhed, containing Diocles' treatise. I also thank the authorities of the Shrine Library, and especially Mr. Ahmad Golchīn-Maʿānī, for their cooperation in providing photographs of the manuscript. Mr. Golchīn-Maʿānī also sent me, most generously, a copy of his catalogue of the astronomical and mathematical manuscripts of the Shrine Library. I am grateful to the Chester Beatty Library, Dublin, and the Universiteits-Bibliotheek, Leiden, for providing me with microfilms of manuscripts I wished to consult, and to the Biblioteca Ambrosiana, Milan, for granting me access to its manuscripts.

The text pages in Arabic script and the Index of Technical Terms were set by a computer-assisted phototypesetting system, using computer programs developed at the University of Washington and a high-speed image-generation phototypesetting device. A continuous stream of text on punched cards was fed through the Katib formatting program, which broke up the text into lines and pages and arranged the section numbers and apparatus on each page. Output from Katib was fed through the compositor program Hattat to create a magnetic tape for use on the VideoComp phototypesetter. The Arabic font was designed by Walter Andrews and Pierre MacKay of the University of Washington. The programs were written by Pierre MacKay, and the final sheets were produced on the VideoComp operated by Arcata Graphics, Data/Composition Operations, in San Francisco. I am especially grateful to Pierre MacKay for placing this system at my disposal and for the time and trouble he took adapting his programs to the special requirements of this text. I also thank my colleague, Charles Strauss, for instructing me in the use of the IBM CP-67/CMS Editor.

I thank my colleague, W.O. Beeman, for help in reading and interpreting the Persian text on p. 114; Ethel Eaton for lending me her expertise in ancient technology and its bibliography; my colleague, David Pingree, for information on Indian matters; Prof. Franz Rosenthal, for giving me the correct reading of the colophon of the Meshhed ms.; Janet Sachs, for help in typing a difficult manuscript; Jonathan Sachs, for drawing most of the figures; my pupil, Jacques Sesiano, for suggesting a number of improvements in the text and its interpretation; and Prof. Dr. Fuat Sezgin, for his constant willingness to share his unique knowledge of Islamic manuscripts, and particularly because he drew my attention to the existence of the Meshhed ms. and was instrumental in obtaining photographs of it.

I am as always, grateful to my teacher and colleague, O. Neugebauer, for constant help and advice. I am particularly glad to be allowed to publish two contributions by him as Appendices C and D. Finally, I wish to record my debt to S. M. Stern, one of the finest scholars I have ever known, whose premature death was a great loss, not only to Islamic studies, but to scholarship as a whole. He first drew my attention to the existence of Diocles' treatise, encouraged me to edit it, and was generous with his expert help. Were he still living, this would be a better book. Such as it is, I dedicate it to his memory.

Providence, December 1975 G. J. Toomer

Table of Contents

Introduction

1. Life of Diocles

We know nothing about Diocles except what can be inferred from the present work. Until the discovery of the Arabic text, even his date was in doubt. The extracts from his work transmitted by Eutocius (see p. 18) show that he must have lived after Archimedes, since he supplies a solution to a problem left incomplete in the text of Archimedes [1]. Unfortunately Eutocius also inserted references to Apollonius' *Conics* in his version of Diocles [2]. Despite Tannery's warning that these must be additions by Eutocius [3], they were used to establish a *terminus a quo*. A *terminus ad quem* was devised from the name "cissoid", which was used by Geminus. On the assumption that it was the ancient name for the curve described by Diocles [4], it was concluded that he lived before Geminus. Thus Heath, HGM I p. 264, asserts that Diocles lived after Archimedes and Apollonius and before Geminus, i. e. (according to Heath's erroneous datings of Apollonius and Geminus) between 200 and 70 B.C. [5].

The full text of Diocles' treatise allows us to establish his date with some precision. It confirms Tannery's suspicion that the references to Apollonius are additions by Eutocius. With one

---

1 Eutocius, Heiberg III pp.160-62. In our text, §§136-49
2 Heiberg III p.168, 10-12; p.170, 16-17 and 22-23.
3 Tannery, "La cissoide" p.46.
4 §219 ff. Eutocius, Heiberg III p.68. The assumption is in fact extremely dubious: see p. 24.
5 There are no grounds for Heath's further refinements, "towards the end of the second century" (ibid.), "a century or more later than Apollonius" (ibid. II p.200). For the use of Geminus as *terminus ad quem* see also Cantor, *Vorlesungen* p.354. On the probable date of Geminus (first century A.D. rather than the conventional first century B.C.) see Neugebauer, *History of Astronomy* pp.579-80.

exception, Diocles refers only to mathematicians of Archimedes' generation or the preceding one, Conon, Pythion, and Dositheus [6], i. e. men who flourished about the middle of the third century B. C. The exception is Zenodorus. Although the Arabic text is slightly corrupt at both places where this person is mentioned, that is the only plausible way to read the name. Since Zenodorus was in personal contact with Diocles, and can himself be dated by his association with the philosopher Philonides, we can assign the "floruit" of Diocles with assurance to the early second century B.C. (roughly 190-180) [7]. This makes him an exact contemporary of Apollonius (who was himself acquainted with the young Philonides) [8]. It also makes him close in time to Dionysodorus, who, like Diocles, supplied a solution to Archimedes' problem, and who is named as a teacher of Philonides [9]. This external dating fits well with the place we should assign Diocles in the history of the theory of conics (essentially pre-Apollonian) from analysis of the present work (see pp. 9-17).

The only other information on Diocles' life afforded by our text is that when Zenodorus propounded to him the query from which the book starts he was living in Arcadia (§4). It would be wrong to conclude from this that Arcadia was a "cultural center" in this period (though the future historian Polybius was growing up at Megalepolis at about this time): the whole of the introduction (§§3-7) confirms the impression we derive from other contemporary sources, that mathematics during the Hellenistic period was pursued, not in "schools" established in "cultural centers", but by individuals all over the Greek world, who were in lively communication with each other both by correspondence and in their travels [10].

---

6 See §§ 3,6,136,149, with notes ad locc.

7 I omit the detailed evidence for the dating of Zenodorus, since I have discussed it exhaustively in my article "The Mathematican Zenodorus". I suggest there that he is to be identified with a Zenodorus who appears in an Athenian inscription of 183/2 B.C., but the dating in no way depends on that identification.

8 Apollonius, *Conics* II Introduction, Heiberg p.192,8-11. The evidence for dating Apollonius' mathematical activity in the late third and early second centuries B.C. is summarized in my article "Apollonius of Perga" pp.179-80,192. It is conveniently given *in extenso* by P.Fraser, *Ptolemaic Alexandria* II pp.600-03 (notes 316-26). Fraser's own discussion (ibid. I p.415) is worthless.

9 Crönert pp.945,952,956. For what is known about Dionysodorus see Bulmer-Thomas, "Dionysodorus". His solution to Archimedes' problem is given by Eutocius, Heiberg III pp.152-60.

10 The failure of P.Fraser to understand this vitiates the whole section (Ch.VII Pt.II) on mathematics of his *Ptolemaic Alexandria*.

2. Diocles' Work: Title and Subject

The Arabic title, *On Burning Mirrors* (fī ʾl-marāyā ʾl-muḥriqa), is a correct translation of its Greek predecessor, περὶ πυρίων, which is the form in which it is quoted by Eutocius [1]. Whether it was given this title by Diocles himself is dubious. It is true that the treatise starts from two problems concerning burning-mirrors. These are answered in Prop. 1, which deals with the parabolic mirror. Props. 2 and 3 deal with the spherical burning mirror, Props. 4 and 5 again with the parabolic mirror. But the rest of the treatise has nothing to do with burning-mirrors. If we ignore the spurious Props. 6 and 9 (see pp.161-2 and 168), then Props. 7 and 8 deal with a problem left unsolved in Archimedes' *Sphere and Cylinder*, and Props. 10-16 with the problem of doubling the cube. One can trace a certain logical sequence in the propositions. The problems with which Diocles starts require theorems in conics for their solution. After dealing with burning-mirrors, he then proceeds to another problem requiring conics (Archimedes'), then another (doubling the cube, which he solves by the intersection of two parabolas). Then he propounds another solution of doubling the cube, this time using not conics but a special curve (the cissoid). However, it must be admitted that the connection is tenuous, and the work is in reality a collection of groups of theorems in higher geometry which have little in common but their author. It is conceivable, as J. Sesiano suggested to me, that what we have is three separate short works (on burning-mirrors, Archimedes' problem, and doubling the cube) which were combined into one in the course of transmission. If that is so, the combination had already taken place by the time of Eutocius (sixth century A. D.).

3. The Theory of Conic Sections up to the Time of Diocles

Much of the mathematical part of *On Burning Mirrors* employs theorems in conics. To appreciate Diocles' own contribution, one must know something of the state of the theory of conic sections when he wrote the treatise. This, however, is a matter of considerable uncertainty. The only systematic treatise on the theory surviving from

1 Heiberg III pp.66,8; 130,23; 160,3,4. The form of the word in Diocles' time was presumably still πυρεῖον, but the confusion between ει and ι is common (e.g. in inscriptions) in the Roman period. The form πυρῖον is found in Anthemius and the Bobbio Mathematical Fragment (*Mathematici Graeci Minores* pp.85,9; 88,12).

antiquity is Apollonius' *Conics* (what knowledge, if any, Diocles had of that treatise is a delicate question, as will be seen). Besides that, we have only some works of Archimedes *(Conoids and Spheroids, Equilibriums of Planes, Quadrature of the Parabola, On Floating Bodies, The Method)* which make some highly specific applications of conics and incidentally allow a few inferences about the existence in his time of theorems in elementary conics[1]. Archimedes is also our main (and most reliable) source for pre-Apollonian terminology in conics. Otherwise we have only some lemmas to earlier works on conics in Book VII of Pappus' *Collection*, and some sketchy accounts of some parts of the early history of conics in late authors, notably Pappus and Eutocius. These provide only second- or third-hand information, and are suspect in many ways. Nevertheless, from them and from the surviving texts the following conventional account has been developed in modern times[2]. (We shall see later what features in the account appear difficult to sustain in the light of Diocles' treatise).

The theory of conic sections was invented by Menaechmus (mid-fourth century B. C.) The three sections were obtained by cutting a right circular cone by a plane at right angles to a generator. If the cone is right-angled this produces a parabola, if obtuse-angled a hyperbola, if acute-angled an ellipse. The three sections were accordingly named "section of a right-angled cone", "section of an obtuse-angled cone" and "section of an acute-angled cone" respectively. These are the names still applied to them by Archimedes, more than a century after Menaechmus[3].

---

1 For examples see notes on §§40,41,170,171. The best account of what can be inferred from Archimedes about the theory of conics in his time is still Heiberg, "Kenntnisse des Archimedes", though it needs supplementing.

2 Best in Zeuthen, *Kegelschnitte* (for some important idiosyncrasies of Zeuthen see pp.13,16). Derived almost entirely from Zeuthen (in places word for word) is Heath, *Apollonius* pp.xvii-lxxxvi, but it is a convenient collection of the scattered facts. Essentially the same in Toomer, "Apollonius" pp.180-85, which also owes much to Dijksterhuis, *Archimedes*, particularly pp.55-79.

3 The occurrence of the word παραβολή in Archimedes' *Method* (Heiberg II p.436,1; p.498,32) is considered, plausibly, to be due to the extant text being a revision of Archimedes' original (all traces of the Syracusan dialect which Archimedes normally used have vanished): see Heiberg, "Eine neue Archimedeshandschrift" pp.297-98. The occurrence of ἔλλειψις in *Conoids and Spheroids*, Heiberg I p.292,9; p.298,26 and p.300,7, is certainly due to interpolation. See Heiberg, "Kenntnisse" pp.43-44.

Between the time of Menaechmus and Archimedes treatises on conics were written by Aristaeus and Euclid. We have almost no direct information about the contents of these. However, Archimedes refers to certain theorems as proved "in the elements of conics" (ἔν τοῖς κωνικοῖς στοιχείοις)[4]. This is usually taken to refer to one or both of the works of Aristaeus and Euclid[5]. I prefer to regard it as a vague rather than a specific reference, to "elementary works on conics" (in much the same way as Archimedes means by ἔν τῇ στοιχειώσει[6] not, as is commonly supposed, a specific work by a specific man, namely Euclid, but "the accepted body of theorems in elementary geometry"). Whichever view is correct, by the time of Archimedes there existed a body of theorems on conic sections. The content of some of these can be inferred from Archimedes' works.

I will not list such theorems here (for an attempt to do so see p. 4 n. 1). Instead, I will describe those features in the pre-Apollonian theory of conics which seem to distinguish it from the classical, Apollonian theory. The most obvious is the way of generating the curves, which is reflected in the nomenclature ("section of a right-angled cone", etc.) With this method of generation, each of the three curves can be characterized by what we may call a "symptoma" (adopting the Greek term σύμπτωμα, a constant relationship between certain magnitudes which vary according to the position of an arbitrary point on the curve; a symptoma sometimes, but not always, corresponds to the modern "equation of the curve"). Consider Fig. III (p. 11), which represents a right-angled cone cut by a plane perpendicular to a generator, which produces a parabola with vertex Z and axis ZF. For an arbitrary point K one can prove that

$$KL^2 = 2AZ.\,ZL \qquad (1)$$

(for a proof see p.10). In algebraic notation, if KL = y, ZL = x, 2AZ = p,

$$y^2 = px. \qquad (1a)$$

---

4 *Quadrature of the Parabola* III, Heiberg II p.268,3. Cf. *Conoids and Spheroids* III, Heiberg I p.274,3 and *Floating Bodies* II,2, Heiberg II p.350,8-9.

5 E.g. Heiberg, *Archimedes* II p.269 n.2; Heath, *Apollonius* p.xxxv.

6 *Sphere and Cylinder* I 6, Heiberg I p.20,15.

We find Archimedes using this relationship in the parabola, and calling p (the modern parameter) "the double of the distance to the axis" (ἁ διπλασία τᾶς μέχρι τοῦ ἄξονος) [7], which can be taken as denoting 2ZA in Fig. III [8]. Similarly for the hyperbola and ellipse (see Figs. IV and V, p.12), we can show that

$$\frac{KL^2}{ZL.PL} = \frac{2ZF}{PZ} \tag{2}$$

(for a proof see p.11). In algebraic notation, if $KL = y$, $ZL = x_1$, $PL = x_2$, $2ZF = p$, $PZ = a$,

$$\frac{y^2}{x_1 x_2} = \frac{p}{a} = \text{constant.} \tag{2a}$$

We do not find this explicitly in Archimedes (for a possible exception see p.13), but we do find him using for both hyperbola and ellipse the equivalent of the relationship

$$\frac{y^2}{x_1 x_2} = \frac{y'^2}{x'_1 x'_2} \quad , \quad ^{9)}$$

and there can be no doubt that the relationship (2) was as well known in the "elements of conics" as was (1). It seems to me virtually certain that by Archimedes' time the relationships (1) and (2) were considered to be the *defining properties* of the three curves (see further pp. 10-15).

The most characteristic feature of this method of defining the curves is that they are in "orthogonal conjugation", i.e. ZL always lies on the axis of the curve and KL is at right angles to the axis. This is reflected in Archimedes' terminology: he calls ZL the "diameter" (διάμετρος), and not the "axis" (ἄξων). A "diameter" of the parabola in the modern (i.e. Apollonian) sense he calls "parallel to the diameter" [10].

---

7 *Conoids and Spheroids* III, Heiberg I p.272,17. For other examples see Heiberg's Index I s.v. μέχρι.

8 For another way of interpreting the phrase see p.13 and note on §38.

9 E.g. for hyperbola *Conoids and Spheroids* XXV, Heiberg I p.376,19-23; for ellipse ibid. VIII, p.294,22-26. For other references see Heiberg, "Kenntnisse" pp.48,55.

10 E.g. *Quadrature of the Parabola I*, Heiberg II p.266,7. For other peculiarities of Archimedes' terminology see Heath, *Apollonius* p.xlix.

Apollonius introduced a new method of generating the curves, by cutting the cone, defined in a much more general form, namely the double oblique circular, by a plane. According to the different dispositions of the cutting plane, the three curves can all be generated from the same cone [11]. Apollonius found symptomata for all three curves, and defined them by the method of "application of areas", which was the standard Greek procedure for formulating geometrically problems which are, algebraically, equations of the second degree [12]. In the parabola, if the ordinate is y and the abscissa x, he represented the symptoma corresponding to equation (1a), $y^2 = px$, by saying that the rectangle of side x and area equal to y^2 is applied (παραβάλλεται) to the line-length p. In the case of hyperbola and ellipse, since PZ = PL - ZL and PL + ZL respectively, Apollonius' equivalent of equation (2a) can be transformed into the equivalent of

$$\frac{y^2}{x(a \pm x)} = \frac{p}{a}$$

(setting ZL = x instead of x_1). Hence

$$y^2 = x(p + \frac{p}{a}x) \text{ for the hyperbola} \qquad (3)$$

$$y^2 = x(p - \frac{p}{a}x) \text{ for the ellipse.} \qquad (4)$$

Apollonius represents the relationship (3) by saying (see Fig. I) that a rectangle of side x and area equal to y^2 is applied to p so that it exceeds it (ὑπερβάλλει) by a rectangle similar to $\frac{p}{a}$. Similarly he represents (4) by saying (see Fig. II) that a rectangle of side x and area equal to y^2 is applied to p so that it falls short of it (ἐλλείπει) by a rectangle similar to $\frac{p}{a}$. Hence he gives the curves the names "parabola", "hyperbola" and "ellipse" respectively. The parameter p he calls ὀρθία, i. e. the side of the applied rectangle which is *perpendicular* to the ordinate. He also calls it "[the line] to which [when there is applied a rectangle with side equal to the abscissa] the ordinates are equal in square [to that rectangle]" [13].

---

11 For details see e.g. Toomer, "Apollonius" pp.181-85.

12 The *locus classicus* is Euclid VI 28-29. The method is closely related to the "geometrical algebra" of Euclid II.

13 παρ' ἣν δύνανται αἱ καταγόμεναι τεταγμένως. I expand this highly abbreviated expression to παρ' ἣν [παραβαλλόμενον ὀρθογώνιον, οὗ ἡ πλαγία πλευρὰ ἴση τῇ ἀπολαμβανομένῃ ὑπ' αὐτῆς] δύναται ἡ κατηγμένη τεταγμένως. (Changed to the plural because it is true of *every* ordinate). Similarly Mugler, *Dictionnaire* p.151.

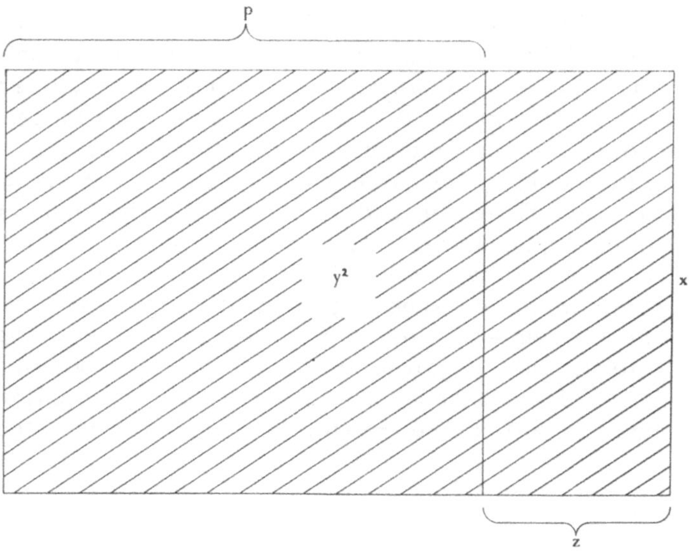

Fig. I

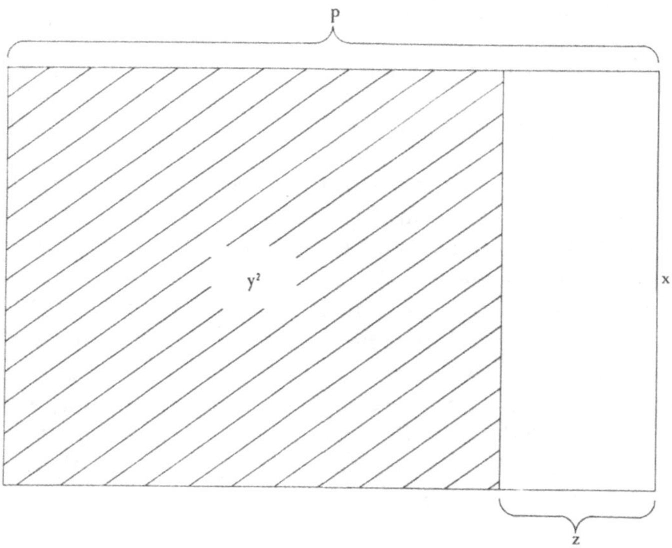

Fig. II

Apollonius' method of generating the curves differs in another significant respect from the older method: it produces immediately oblique conjugation. That is, the relationships (1) to (4) no longer apply to the axis of the curve and orthogonal ordinates, but to an arbitrary diameter and the conjugate ordinates (i. e. the parallels to the tangent at that diameter). Apollonius proves at an early stage in the *Conics* (I 50 and preceding propositions) that one can establish relationships equivalent to (1), (3) and (4) for *any* diameter and the conjugate ordinates.

4. Conics in *On Burning Mirrors*

(i) *Theory and terminology*. If we now examine Diocles' treatise in the light of the above conventional account, we find that it would fit in with it very well, if we could only ignore Prop. 8. Apart from that proposition, it reads very much as we would expect a work written in the pre-Apollonian tradition to do. Diocles consistently uses the term "section of a right-angled cone" for the parabola (see note on §8), and he treats it only in orthogonal conjugation. He represents *half* the parameter of the parabola as a line-length perpendicular to the axis of the parabola, in exact accordance with the older definition (see note on §38). He assumes without proof the theorem that the subnormal in the parabola is constant and equal to the half-parameter. This theorem does *not* appear in Apollonius' *Conics*, but we have independent evidence that it was a theorem in the pre-Apollonian "elements of conics" (see note on §41). Diocles uses the term ἄξων only for the axis of a conoid, never for the axis of a conic (see note on §9, p. 142). For the latter he uses "the bisector", which accords with neither Archimedes' nor Apollonius' practice (see note on §8, p. 141). He does indeed use the Apollonian term ἡ παρ' ἣν δύνανται for the parameter, but since this is also found once in Archimedes no conclusion can be drawn (see note on §9, p. 141).

The external evidence, as we have seen (pp. 1-2), makes Diocles an exact contemporary of Apollonius, so it would hardly be surprising if he showed no knowledge of his *Conics*. But Prop. 8 poses a difficult problem. There alone in the whole work appear hyperbola and ellipse. Diocles calls them not, as one would expect from his nomenclature for the parabola, "section of an obtuse-angled (acute-angled) cone", but ὑπερβολή and ἔλλειψις, i. e. the "Apollonian" names. Furthermore, the symptoma of the ellipse is applied in oblique conjugation. Now it is not plausible to suppose that either in the Greek or the Arabic transmission of the text someone altered the terminology here and here alone. We must therefore recognize that Diocles himself used the terms, and is responsible for the inconsistent terminology. No explanation of this inconsistency can be anything more than a suggestion. It is possible that Diocles acquired a copy of Apollonius' *Conics* soon after it was published[1] and adopted the new terminology in this one passage. But we should also consider the possibility that the conventional account of the history of conics is not altogether correct.

To explain what I mean, I revert to the "pre-Apollonian" generation of the three curves from the three different types of cone. These are depicted in Figs. III-V (right-angled, obtuse and acute respectively). In each case the axis of the cone is AF and the cutting plane is ZKL, Z being the vertex of the section, K an arbitrary point on the curve, and L the point where the perpendicular from K meets the axis of the section ZF. MKN is the circular section of the cone through K, MLN the diameter of that circle passing through L. ZG is the diameter of the circular section of the cone through Z parallel to MN. GH and NE are drawn parallel to AF. In Fig. IV NGA is produced to meet FZ (produced) in P. Similarly in Fig. V AGN is produced to meet ZF (produced) in P. Then in every case

$$KL^2 = ML.LN \quad \text{(Euclid III 35)}$$
$$\text{and } ML.LN = ZL.LE \quad \text{(ZML, NEL similar triangles).}$$

Then for the parabola (Fig. III)

$$\frac{EL}{HZ} = \frac{NL}{GZ} \quad \text{(similar triangles)}$$
$$\text{but } NL = GZ \quad (AN \parallel ZF)$$
$$\therefore \quad EL = HZ = 2ZF$$
$$\therefore KL^2 = ZL.2ZF \qquad (1)$$

or, since this is a right-angled cone, and $\widehat{ZAF} = 45° = \widehat{ZFA}$,

$$KL^2 = ZL.2ZA.$$

---

1 We learn from Apollonius' preface to Bk.I of the *Conics* (Heiberg p.2,18-21) that copies of the first two books were in circulation before he began to "publish" the work. It is useless to speculate about the probability of Diocles having known the *Conics* when we are ignorant of most of the relevant facts: the precise dates of publication of the *Conics* (we know only that it was late in Apollonius' career) and of *On Burning Mirrors;* what "publication" of a book really meant in the Hellenistic period (multiplication of copies, circulation, etc.); whether Apollonius and Diocles were personally acquainted.

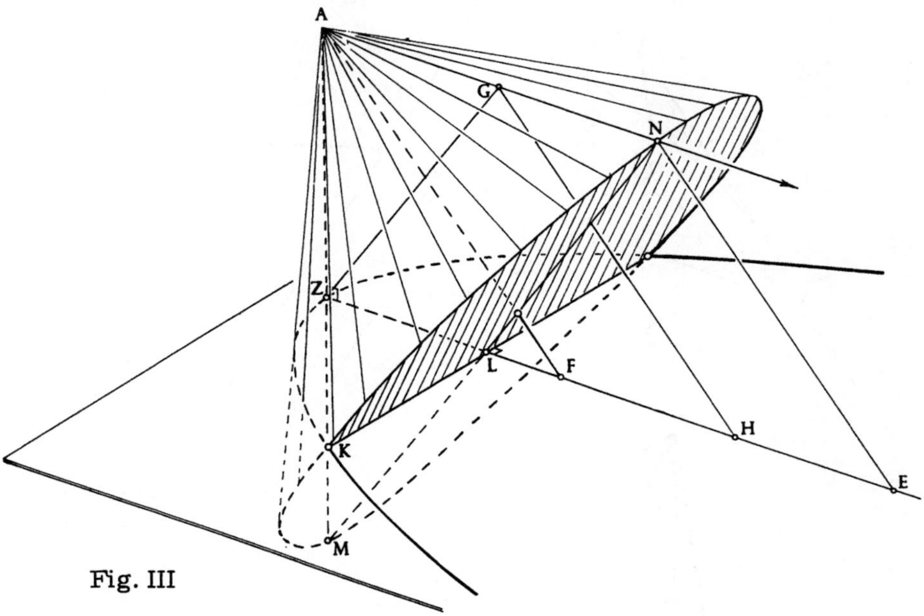

Fig. III

For the hyperbola and ellipse (Figs. IV and V)

$$\frac{EL}{HZ} = \frac{NL}{GZ} = \frac{PL}{PZ} \quad \text{(similar triangles)}$$

$$\therefore \frac{EL}{PL} = \frac{HZ}{PZ}$$

$$\therefore \frac{ZL.LE}{ZL.PL} = \frac{HZ}{PZ} = \frac{2ZF}{PZ}$$

$$\therefore \frac{KL^2}{ZL.PL} = \frac{2ZF}{PZ}. \quad 2) \qquad (2)$$

Now, for an arbitrary point K, KL is the ordinate and ZL the abcissa. 2ZF is a constant, twice the distance from the vertex of the section

---

2 These proofs of the fundamental properties are close to or identical with those suggested by Dijksterhuis, *Archimedes* pp.58-59.

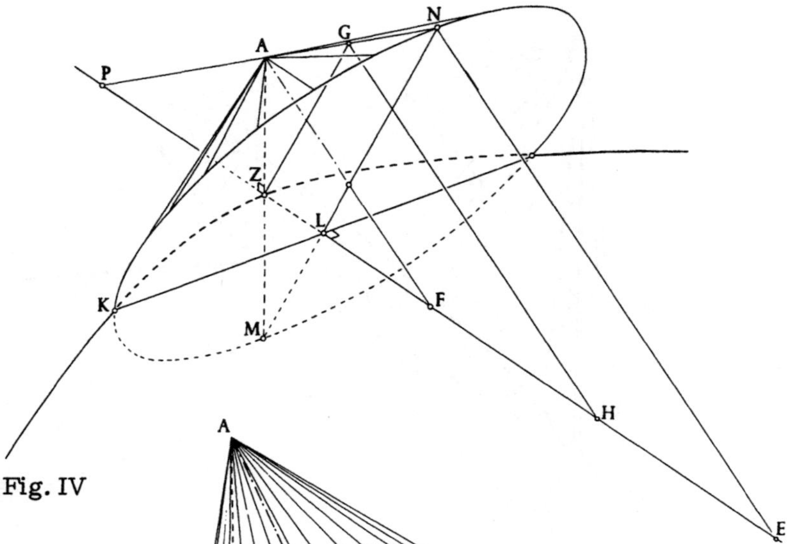

Fig. IV

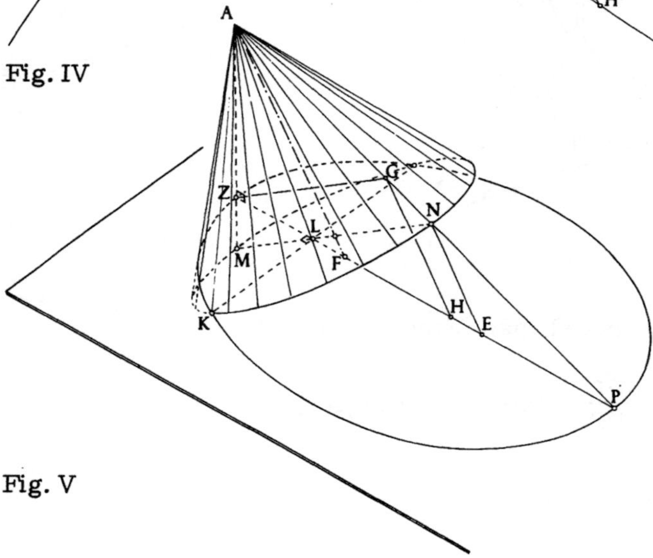

Fig. V

to the axis of the cone (measured along the axis of the section). PZ is another constant (the transverse diameter in the hyperbola, and the major axis in the ellipse) and PL = PZ ± ZL (in hyperbola and ellipse respectively). Thus (2) can be reformulated as

$$KL^2 = ZL\,(2ZF + \frac{2ZF.\,ZL}{PZ}) \quad \text{for the hyperbola}$$

and $KL^2 = ZL\,(2ZF - \frac{2ZF.\,ZL}{PZ})$ for the ellipse, which are the formulations (3) and (4) on p. 7:

$$y^2 = x\,(p + \frac{p}{a}x)$$

$$\text{and } y^2 = x\,(p - \frac{p}{a}x) \quad .$$

These can be represented by the method of "application of areas" exactly as Apollonius did (see p. 7). The only difference is that here we are using orthogonal conjugation (so KL is always perpendicular to ZL) and the parameter p, which for Apollonius is a rather complicated ratio between different constants in his generating figure, is here simply twice an actual length in the curve, the distance from the vertex to the cone's axis.

Not only is it possible to express the symptomata of the sections, generated in the old fashion, in terms of application of areas,[3] but, in Greek geometry, it is natural to do so. I do not doubt that this was done at an early stage (certainly before Archimedes), and I am glad to find that I have reached the same conclusion as Zeuthen[4]. It is interesting that we find the phrase "the double of the distance to the axis" (ἁ διπλασία τᾶς μέχρι τοῦ ἄξονος) used for the parameter of the parabola by Archimedes[5]. If I am right, it meant "the parameter" (for all three curves) in early conics[6]. Furthermore, there is one passage in Archimedes which actually refers to the use of the applied rectangle in the case of the hyperbola[7]. It is true that Heiberg considers this an interpolation, and he may well be right, since the passage is difficult

---

3 My statement, "Apollonius" p.184, that the older approach allowed this to be done only for the parabola, is simply false.

4 *Kegelschnitte* pp.55-56. His arguments are reproduced, but dismissed, by Heath, *Apollonius* pp.lxxxi-iv. No-one seems to have followed Zeuthen in this.

5 See p.6 n.7. In the case of the parabola the phrase is ambiguous: it can refer to either 2ZF or 2ZA in Fig.III (see p.11), and either can be plausibly interpreted as "the distance to the axis", ZF as the distance along the axis of the parabola, ZA as the distance along the surface of the cone. There is support for the latter interpretation in the way Diocles draws his Fig.1; see note on §38. But consistency would require us to interpret it in the former way, since nothing else will fit ellipse and hyperbola. I do not see how to resolve this dilemma.

6 Cf. Zeuthen, *Kegelschnitte* pp.465-66.

7 *Conoids and Spheroids* XXV, Heiberg I p.376,22-25.

to interpret in a way that makes sense mathematically, but his argument, that this is an Apollonian concept, is circular.

Now if, as I have argued, the symptomata of the three sections were defined by the application of areas under the old system of generation, the applied rectangles for the "section of an obtuse-angled cone" and the "section of an acute-angled cone" necessarily "exceeded" and "fell short of" the parameter respectively. Hence the names "hyperbola" and "ellipse" could have been applied to the curves *before* Apollonius introduced his new system of generation, and are not necessarily tied to the latter. It has always been taken for granted that the names "parabola", "hyperbola" and "ellipse" were *introduced* by Apollonius. The reasons are, first, that Pappus says so[8], and second, that Apollonius, in defining the curves, says "let it be called (καλείσθω) parabola" etc.[9] Neither argument is conclusive. Though much better informed than we are about earlier Greek geometry, Pappus is often careless and sometimes demonstrably in error. The use of καλείσθω does not necessarily imply that the term had never been used in this sense before. A counterexample is found in the line of Apollonius' text immediately following the definition of the parabola referred to above: Apollonius says, let ΘZ (the parameter) be called παρ' ἣν δύνανται etc., a phrase, which, as we have seen (p. 9), is already found in Archimedes. It is at least conceivable, then, that the "area application" nomenclature for the sections was coined before Apollonius, and was one of a number of competing systems in use at the time Diocles wrote his treatise[10]. In that case it was canonized, rather than created, by Apollonius. We might even conjecture that only the terms "hyperbola" and "ellipse" were originally coined, and that the less obvious "parabola" was devised by analogy later, perhaps by Apollonius himself (it is less obviously appropriate because there is an "application", παραβολή, in all three cases). Though disinclined to accept such an *ad hoc* hypothesis, I consider it possible that the "area

---

8 *Collection*, Hultsch p.674,5-7. A similar distinction between the old and new nomenclatures is made by Eutocius in his commentary on the *Conics* (Heiberg II pp.168-74), allegedly drawing on Geminus. But Eutocius does not explicitly attribute the terms "parabola" etc. to Apollonius. His explanations of why they are so called (ibid. 172-74) are notoriously absurd.

9 E.g. *Conics* I 11, Heiberg p.42,1.

10 For the ellipse we also find the term θυρεός (shield) at an early period, e.g. Proclus, *Comm. on Euclid I*, Friedlein p.126,19; cf. Euclid, *Phaenomena*, Menge p.6,7.

application" nomenclature for the sections, as well as the procedure it implies, long precede Apollonius. The fact that Archimedes uses the older nomenclature is no proof that the other did not exist in his time: it merely shows that it had not yet ousted its predecessor. But however one answers the questions raised by the new evidence difficulties remain.

The other anomalous feature of Prop. 8, the use of the symptoma of the ellipse in oblique conjugation, is less difficult to explain. It is obvious to anyone who reads Archimedes carefully that he was fully aware that the defining symptomata of the sections (in orthogonal conjugation) had their analogies in oblique conjugation [11]. I have no doubt that this too was part of "the elements of conics". Apollonius' contribution was not to discover the properties in oblique conjugation, but to introduce them immediately as *defining* properties by the greater generality of his method of generation [12].

(ii) The focus of the parabola. In Bk. III Props. 45-52 of the *Conics* Apollonius deals with certain properties of the foci [13] of ellipse and hyperbola. It has long been a puzzle why he never mentions the focus of the parabola. Some have concluded that he was simply unaware of its existence [14], but all who have read the *Conics* with any care agree that he must have known of it [15]. Before the discovery of the present treatise, the earliest surviving examples of the use of the focal property of the parabola (i. e. reflection of parallel rays to the focus) were from late antiquity, the Bobbio Mathematical Fragment and Anthemius (sixth century A.D.) [16]. Now, however, we know that that property was

---

11 For some evidence see my note on §170; see further Dijksterhuis, *Archimedes* pp.66 n.1, 106; Toomer, "Apollonius" p.186.

12 Exactly as he himself says, *Conics* I Introduction (Heiberg p.4,1-4): "Book I contains...the basic symptomata [of the sections] worked out more fully and generally than in the writings of others".

13 The term "focus" for these points was introduced by Kepler in his work of 1604, *Ad Vitellionem Paralipomena* IV 4, *Werke* 2 p.91. I know of no ancient or medieval term. Apollonius refers to them vaguely as "the points arising from the application" (τὰ ἐκ τῆς παραβολῆς γενηθέντα σημεῖα), e.g. Heiberg p.424,10-12.

14 E.g. Cantor, *Vorlesungen* I p.339 (more cautiously ibid. p.344).

15 E.g. Zeuthen, *Kegelschnitte* pp.367-73. Neugebauer, "Apollonius-Studien" pp.236-42, attempts to show how the focus of the parabola and the focal property could be derived in a way analogous to Apollonius' procedure for the foci of ellipse and hyperbola.

16 Heiberg, *Mathematici Graeci Minores* pp.85-86,87-88. A paraphrase of the Bobbio version is given in Appendix B(i). An inaccurate English translation of both works was published by Huxley, *Anthemius of Tralles*.

recognized long before Apollonius, since Diocles informs us that the problem of constructing a burning-mirror which makes all the rays meet in one point was solved by Dositheus [17], i. e. in the mid-third century. Unfortunately it is unclear from the Arabic text precisely what Dositheus did, and Diocles seems to imply that no one before himself had given a formal geometric proof of the focal property of the parabola. But at the very least Dositheus must have enunciated that property. Thus we can be sure that Apollonius was well aware of it, and though we still do not know why he omitted all mention of the focus of the parabola in his *Conics*, we know that he did it deliberately. The reason may well be that he had already discussed the subject in another work. I suggested ("Apollonius" p. 187) that he did so in the work *On the Burning-Mirror* (περὶ τοῦ πυρίου) ascribed to him in the Bobbio Mathematical Fragment [18]. I now believe, however, that the author of the fragment is referring to none other than the present work of Diocles, which he mistakenly attributes to Apollonius (see p. 20). Zeuthen suggested that Apollonius treated the subject in his lost work *Tangencies* [19]. Another possibility is that he knew the treatment by Diocles and did not wish to repeat it.

We now know, too, that Anthemius' statement that "the ancients" indeed constructed burning-mirrors, but gave no geometrical proofs for their construction [20], is worthless, like much else in his treatise. I omit mention of modern discussions of this point, since the present publication makes them obsolete; I merely note that the more perceptive [21] correctly inferred from the passage of al-Akfānī translated by Wiedemann (see p. 21) that Diocles gave a mathematical treatment of at least the parabolic burning-mirror.

(iii) Construction of the parabola from focus and directrix. In Prop. 4 Diocles solves the problem "to construct a burning-mirror of given focal length" by a method which is essentially drawing a parabola by means of focus and directrix. In Prop. 10 he uses the same method to construct two parabolas. In previously extant Greek

---

17 §6. See note ad loc. on this man.

18 *Mathematici Graeci Minores* p.88,8-12. Cf. Zeuthen, *Kegelschnitte* pp.378-79, for a similar conclusion.

19 *Kegelschnitte* p.371 n.1. On this work see Heath, HGM II pp.181-85.

20 *Mathematici Graeci Minores* p.85,11-16. Anthemius' own treatise could be criticized on much the same grounds.

21 E.g. Heath, HGM II p.201.

literature such a construction was found only in Anthemius[22]. However, Pappus proves that, given a straight line AB and a fixed point G, the locus of a point D moving such that the ratio of its distance from G and its vertical distance from AB is constant will be a conic, and will be a parabola if the ratio is equal to 1, an ellipse if less than 1, and a hyperbola if greater than 1[23]. Thus the generation of all three sections from focus and directrix was known in the fourth century A.D., and probably much earlier. Indeed, since Pappus gives this theorem as a lemma to Euclid's *Surface Loci*, it has even been argued that Euclid stated it without proof, and that therefore it must have been proved earlier, e. g. by Aristaeus[24]. Since we know virtually nothing about Euclid's *Surface Loci*, this was always a very dubious inference, and it is now, I believe, definitely disproved by the present work. For Diocles, having used the method to draw a parabola in Prop. 4, finds it necessary in Prop. 5 to *prove* that the curve so drawn is indeed a parabola, i. e. that it has the defining symptoma of the parabola. He would not have done this if the focus-directrix property had been a well-known theorem, established a century earlier[25]. We can, then, attribute to Diocles the discovery of the focus-directrix property for the parabola. The extension to all three sections must belong to a later time. I am sure that it was an achievement of the Hellenistic period, and not original with Pappus. An obvious candidate is Apollonius, in one of his lost works on loci, but this is merest conjecture.

---

22 *Mathematici Graeci Minores* pp.85,19 to 87, 3. The work, as we have it, breaks off at this point, so it is impossible to say whether Anthemius ever proved that the points he constructs lie on a parabola. His construction (which is clumsy compared with Diocles' elegant method) does indeed imply the focus-directrix property, but whether Anthemius was really aware of that property seems dubious to me.

23 Pappus, *Collection* VII 312-18, Hultsch pp.1004-14. For the formulation see especially p.1012, 24 to 1014, 2.

24 Heath, HGM II p.119. More cautiously Zeuthen, *Kegelschnitte* p.370. I regret my approval, "Apollonius" p.187.

25 Contrast his use without proof of theorems in conics which were indeed well established, notes on §§40, 41.

5. Influence of *On Burning Mirrors*

There is no reference to Diocles, nor any trace of influence of his work [1], in surviving Greek literature until very late antiquity. In particular, there is nothing in Pappus' *Collection*, the source of most of our knowledge about lost Greek mathematical works, to indicate that he knew *On Burning Mirrors*. It was perhaps mere chance that a copy of the work survived and came into the hands of Eutocius (?at Alexandria) [2]. His extensive quotations from it in his commentary on Archimedes' *Sphere and Cylinder* were the only source of knowledge of it in the West until the discovery of the Arabic translation. Eutocius gives the substance of Props. 7-8, 10 and 11-13 [3]. Unfortunately, except in Prop.7. he does not quote Diocles directly, but reformulates his proofs to conform to the scholastic norms of his own time (essentially "Euclidean" and Apollonian). This has led to serious misconceptions about Diocles and his work [4].

There is no real evidence that Eutocius' friend, Anthemius of Tralles, knew Diocles' work, despite the similarity in content of some of his *On Paradoxical Devices*. But I believe that *On Burning Mirrors* was probably known to the author of another work from about the same time, the "Bobbio Mathematical Fragment". This

---

1 With the exception of the extremely hypothetical influence on Apollonius suggested p.16.

2 We do not know where Eutocius worked. He dedicated his commentary on Archimedes to a philosopher Ammonius (Heiberg III p.2,16), who is probably the well-known man, son of the Hermias who was a fellow-student of Proclus. Since Ammonius taught at Alexandria, it is likely that Eutocius was there for at least part of his life. See Tannery, "Eutocius", who argues convincingly that Eutocius' working career belongs to the early sixth century. This is confirmed by a horoscope for October 28, 497 which is said (in one manuscript) to be from Eutocius' astrological work. See Neugebauer-Van Hoesen no.L 497, pp.152-57, 188-89. I think it not unlikely that this is in fact Eutocius' own horoscope. From a mention of a discussion by Eutocius on Aristotelian logic in a commentary by Elias to the *Prior Analytics*, Westerink ("Elias" pp.129-31) drew the conclusion that Eutocius occupied the "chair of philosophy" at Alexandria between Ammonius (d.ca.520) and Olympiodorus. This conjecture remains unconfirmed.

3 To facilitate comparison with Diocles' text I give text and translation of Eutocius' excerpts in Appendix A.

4 See e.g. p.1 (Diocles' date) and notes on §§186-207,219. Since Eutocius excerpted Prop.10 without giving the author's name, it has been mistaken for a proof by Menaechmus in modern times.

work has survived in the manuscript Milan, Ambrosian L 99 sup. (now SP II 65), which is a late-eighth century codex of Isidore's *Etymologies*, formerly in the library of the famous monastery of St. Columban at Bobbio. Some sixteen leaves of this are palimpsest. The underlying script is in Greek capitals of the late antiquity (seventh or possibly sixth century). All the palimpsest leaves are from works of mathematical content, including some fragments of Ptolemy's *Analemma*, otherwise known only in William of Moerbeke's Latin translation from the Greek[5]. They also contain parts of an otherwise unknown mathematical treatise, commonly called the "Bobbio Mathematical Fragment"[6]. Most of what can be deciphered of this is on a single corresponding pair of leaves (pp.113-14, 123-4), which for some reason was never written over[7]. I have examined the manuscript myself. The palimpsest pages are all discolored to a dark brown, evidently through the application of chemicals some time ago (I suspect by Angelo Mai, the discoverer of so many palimpsests, who first published some of this text[8]). Although no ultraviolet lamp was available at the Ambrosian when I was there, I doubt very much if anyone would be able to recover, with modern aids, more of the text than Heiberg was able to read eighty years ago. In fact the chemical damage has now made much of what he deciphered of the *Analemma* illegible. Heiberg's text of the mathematical fragment, published in his *Mathematici Graeci Minores* pp. 87-92, is accurate except in some unimportant details.

There have been a number of conjectures about the authorship of this piece. Cantor's absurd suggestion that it was Diocles[9] is disproved, if there were any need, by the present publication. Heiberg suggested that the author was Anthemius, and that the fragment was simply another part of his *On Paradoxical Devices* (of which only a part

---

5 Heiberg printed what he could read in his edition of the *Analemma*, Ptolemy, *Opera Minora* pp.194-216.

6 For a bibliography of editions and studies of this see Huxley, *Anthemius* pp.31-32. On the manuscript see also Lowe, CLA III p.26 no.353.

7 For a facsimile of these two pages see Belger, *Hermes* 16 (opposite p.112). A facsimile of p.124 was published by Mai (see next note) and reproduced by Wattenbach, *Specimina* Pl.VIII.

8 In his *Ulphilae partium ineditarum Sp.cimen*, Milan, 1819 (non vidi). I find that Belger and Heiberg also came to the same conclusion about Mai (Belger, *Hermes* 16 p.264; Heiberg, "Ptolemäus de Analemmate" p.4: "Angelo Mai hat mit seiner Galläpfeltinctur grossen Schaden angerichtet; sie ist jetzt dunkelbraun geworden").

9 "Über das Fragentum Bobiense" p.642.

survives in Greek)[10]. It is true that the bombastic style and mathematical ineptitude of the Bobbio fragment would be well suited to Anthemius. Nevertheless we can be sure that it is not part of "On Paradoxical Devices". For that work was translated into Arabic in its entirety, and is frequently referred to in Islamic sources. I know of no manuscript of the original translation, but there does exist a "revision" of it by ʿUṭārid b. Muḥammad (?early 10th century)[11]. ʿUṭārid had two copies of the translation of Anthemius' work, and, being unable to understand them, undertook to "emend" it into an intelligible treatise. Since he was an even greater fool than his source, the results are ludicrous. However, behind them one can discern the outlines of Anthemius' treatise, and it is clear that it contained nothing corresponding to the Bobbio fragment. We can also discern some of Anthemius' work, but nothing of the Bobbio fragment, in al-Kindī's *On Rays*[12]. Nevertheless, language and style force us to attribute the Bobbio fragment to the late antiquity (the script of the manuscript forbids a date more recent than the seventh century).

Now the author, in speaking of spherical burning-mirrors, says: "Now the ancients thought that burning took place about the center of the mirror, but Apollonius proved that this was false, very properly [illegible] (?) against the writers on catoptrics, and made clear about what place the burning would occur in his work *On the Burning Mirror*".[13] He goes on to say that Apollonius' proof is too long, and he will provide another. Now the phrase about "the ancients" repeats what Diocles says (§12), and the description of what Apollonius did exactly fits Diocles' achievement in Props. 2 and 3. Rather than supposing that Apollonius went over exactly the same ground as Diocles (or vice versa), I consider it highly probable that the author had the work of Diocles in his hands, and misattributed it to Apollonius (there is no other evidence for a work of this title by Apollonius). This hypothesis would also explain some other peculiarities in the fragment. The author uses the expression "section of a right-angled cone" (as well as "parabola"). He also uses "mixed"

10 Heiberg, *Zum Fragmentum mathematicum Bobiense*, pp.128-29. He is followed by Huxley, *Anthemius* pp.29,32-33.

11 Istanbul, Laleli 2759, 1^v-20^r, "Kitāb ʿamali ʾl-marāyā ʾl-muḥriqa". On the author see Suter no.150 p.67.

12 A facsimile of part of this text has been published by Haschmi, *Propagations of Ray*. It would require closer study than I have given it to determine its relationship to Anthemius.

13 *Mathematici Graeci Minores* p.88,8-12.

angles (between a curve and a straight line). Both these archaic (and un-Apollonian) features could have been borrowed from Diocles [14]. It is true that the proof of the focal property of the parabola given in the fragment [15] differs significantly from Diocles'. But I do not exclude the possibility that the author had access to other older works, nor even that he was capable of (muddled) independent reasoning. In any case, if he did use Diocles' treatise, this would be confirmation that it was circulating in late antiquity.

Further confirmation is provided by the fact that it was translated into Arabic. Generally speaking, the translation of a Greek work into Arabic indicates that it was available, and probably still being read, in the higher schools (especially at Alexandria and Athens) in late antiquity. We have no internal or external evidence about the translator. I can say only that he was obviously thoroughly familiar with standard mathematical terminology, both Greek and Arabic [16]. If forced to hazard a guess, I would suggest Qusṭā b. Lūqā, because of his wide experience as a translator and his known interest in burning-mirrors. But to make this more than an empty conjecture would require a careful and detailed stylistic comparison with known translations of Qusṭā, which I have not undertaken.

Similarly, lack of investigation of the quite considerable surviving Arabic literature on burning-mirrors [17] prevents me from saying much about the influence of Diocles' work in the Islamic mathematical tradition. The only explicit reference to it that I know is in the fourteenth-century encyclopedic work of Muḥammad b. Ibrahīm b. Sāʿid al-Akfānī [18], in a passage to which Wiedemann drew attention long ago [19]: "The ancients used to make these mirrors [i. e. burning-mirrors] out of plane surfaces, but some of them made them concave, until Diocles [20] appeared and proved that when their surfaces are curved in the shape of the parabola, they are of enormous power in burning."

---

14 See §§8 and 81, with notes ad loc.

15 See Appendix B(i) for a paraphrase of this proof.

16 This was by no means always true. For instance, one can infer from the work of ʿUṭārid (see p.20) that the Arabic translation of Anthemius contained some very strange terminology.

17 Wiedemann assembled some bibliographical material on this topic (*Aufsätze* pp.119-20), which is now out of date.

18 Sprenger, *Two works on Arabic Bibliography* pp.14-99. The passage referring to Diocles is on p.82, 12-15.

19 In his *Beiträge zur Geschichte der Naturwissenschaften* V (1905), now conveniently reprinted in his *Aufsätze*, I pp.119-20.

20 "dyūfls", Calcutta text. Corrected by Wiedemann.

Nevertheless, examination of the only Islamic work on the subject available in a modern edition and translation[21], the treatise of ibn al-Haytham on the parabolic burning-mirror, leads me to believe that the author was well acquainted with Diocles' work. Ibn al-Haytham opens with a historical discussion. He says that the ancients (al-mutaqaddimūn) investigated the subject of burning-mirrors. Some constructed them by combining a number of plane or spherical mirrors. These included "Archimedes, Anthemius and others"[22]. Then they (the ancients) discovered the parabolic burning-mirror, but the proofs they gave were unsatisfactory. Ibn al-Haytham proposes to remedy this, which he does by giving a pedantically correct series of proofs in the best Greek scholastic manner (using analysis and synthesis, and enumerating every possible case) of the focal property of the parabola. Ibn al-Haytham, then, knew of a Greek work which treated the parabolic mirror mathematically, but gave "unsatisfactory" proofs. This would exactly describe Prop. 1 of Diocles, which, for one trained in Apollonian conics, is highly unsatisfactory, since it uses without proof a theorem which is not even in Apollonius' *Conics* (see note on § 41). It could also, however, be taken to refer to Anthemius' miserable treatment of the parabolic mirror (see p. 17). What inclines me to the view that ibn al-Haytham is thinking of Diocles is that in his own proof of the focal property his basic theorem is Euclid II 8[23], which is used by Diocles (Prop. 5) to prove that the curve generated from focus and directrix is indeed a parabola (see note on § 118). It is possible that ibn al-Haytham hit on this approach independently, but on balance it seems likely that he was inspired by Diocles[24]. If this is true, Diocles had considerable indirect influence on both

---

21 Most easily in Wiedemann's translation (Heiberg and Wiedemann, "Ibn al Haiṯam" pp.205-18). This work was translated into Latin in the middle ages (Latin text ibid. pp.218-31), and this translation was the main source of discussions of the parabolic burning-mirror in medieval Latin optical works. The Arabic text was printed as the third treatise in ibn al-Haytham's *Majmūʿ al-rasāʾil*.

22 I infer that ibn al-Haytham knew the work of Anthemius (see p.20) and took from it the reference to Archimedes (*Mathematici Graeci Minores* p.85,7-9).

23 See my paraphrase of ibn al-Haytham's proof, Appendix B(ii).

24 Ibn al-Haytham's work on the spherical burning-mirror, however, contains nothing which can be directly related to Diocles' treatment of this topic. This work is printed as the fourth treatise in ibn al-Haytham's *Majmūʿ al-rasāʾil*. A German translation was published by Wiedemann, *Bibliotheca Mathematica* X, pp.293-307.

Islamic and medieval Latin discussions of the parabolic burning-mirror.

Diocles was also known to Islamic authors through the intermediary Eutocius, whose commentary on Archimedes' *Sphere and Cylinder II* was translated into Arabic. This translation is extant in the manuscript Escurial 960 (Casiri 955), 22^v-42^v. It contains Eutocius' extracts from a series of earlier authors (Philon, Heron, Diocles, Menaechmus, etc.) on the problem of doubling the cube. Unfortunately Casiri listed all these extracts in his catalogue as if they were separate works [25], which has led to considerable misunderstanding and confusion in modern bibliographical references. Most modern statements about the existence of an Arabic translation of Diocles' treatise can be traced back to Casiri's erroneous description. The truth was perceived by Heiberg [26], but the error continues to be repeated. References to Diocles' *On Burning Mirrors* in Islamic texts too are sometimes derived from Eutocius. Thus in an extract from the *Zīj al-Ṣafāʾiḥ* of Abū Jaʿfar al-Khāzin (10th century) on the problem of doubling the cube the author says: "What Diocles [27] said about that in his book on burning mirrors", and proceeds to give the solution by means of the cissoid in a version obviously taken from Eutocius rather than from the original [28].

Although I cannot trace any connection with Diocles, I should mention that the problem of Prop. 4 of *On Burning Mirrors*, to construct a parabolic mirror with given focal distance, was solved by Abū'l-Wafāʾ (on whom see pp.29-30) in a most ingenious way in his book *On Geometrical Construction*. The Arabic text has never been published, but three modern versions of it have been printed [29].

The work of Diocles became known in western Europe only through the extracts given by Eutocius. This meant that it was generally unknown until the publication of the *editio princeps* of Archimedes (with Eutocius' commentaries) by Venatorius at Basel

25 Casiri, *Bibliotheca Escurialensis* I p.382. Correctly described in the recent catalogue of Derenbourg-Renaud, p.95.

26 "Zum fragmentum Bobiense" p.128 n.*

27 Corrupted to "nrfly" in ms. Leiden Or.14,296^r, which is my source for the work of al-Khāzin, on whom see Suter no.124.

28 For another fragment of the Arabic translation of Eutocius see Woepcke, *Omar Alkhayyâmî* p.xiii n.**.

29 By Woepcke, "Aboûl Wafâ" pp.325-26 (French translation from a Persian abridgment); by Krasnova, "Abu-l-Vafa" pp.69-70 (Russian translation from the ms. Istanbul, Aya Sofya 2753); and by A.Kubesov, *Al-Farabi* pp.104-06 (Russian translation from the ms. Uppsala, Tornberg 324, where the work is attributed to al-Fārābī).

in 1544. The mathematicans of the late sixteenth and seventeenth centuries devoted much attention, not only to Archimedes, but also to that section of Eutocius' commentary in which he discusses curves used by earlier geometers to solve the problem of doubling the cube. Among these was the cissoid of Diocles. In the seventeenth century the infinite branch of this curve was revealed, and many beautiful properties were discovered by Roberval, Huygens and Newton, among others. For the details of these, and a discussion of the mathematical properties of the cissoid in modern terms, I merely refer to the excellent treatments by Gomes Teixeira and Loria [30]. However, I will mention one problem which has never been adequately discussed, the origin of the name "cissoid" as applied to Diocles' curve.

"Cissoid" is simply a transcription of the Greek κισσοειδής, which means "ivy-shaped". The term is applied bv Pappus [31] and Proclus [32] to a curve or class of curves. The passages in Pappus tell us very little, except that they suggest that the name was given to a class of curves rather than a single curve. From Proclus we learn that it was a closed curve [33], and that it was so named because it came to a point like an ivy-leaf, and thus made an angle with itself. Now the cissoid (in the modern sense) does indeed have a singular point: if in Fig. 13 (p. 100) we draw the branch DΘ corresponding to ZΘ, the curve comes to a point and makes an angle with itself at Θ. But the curve is in no sense a closed curve. It has been suggested [35] that the top half of the generating circle (DHZ in Fig. 13) was counted as part of the curve, the combined figure resembling an ivy leaf. I find this incredible. It is certainly mathematically absurd. We now know that Diocles himself did not call the curve "cissoid". More significantly, neither does Eutocius, although the name was already used by Geminus (first century A.D.), according to Proclus. I therefore consider it in the highest degree unlikely that the name "cissoid" was ever applied in antiquity to Diocles' curve. Nevertheless, it is always so named from the early seventeenth century on, with hardly an indication that the nomenclature rests on a modern conjecture [36].

---

30 Gomes Teixeira I pp.1-26; Loria, *Spezielle Kurven* I pp.36-51.

31 *Collection* III 20 and IV 58, Hultsch I p.54, 21 and p.270, 27-28.

32 *Comm. in Euclid.* (see references in Friedlein's index, p.475).

33 Ibid. p.152, 7-9, cf.111, 5-6; 187, 19-21.

34 Ibid. p.126, 24-26.

35 E.g. Loria, *Spezielle Kurven* p.37.

36 Loria (ibid.) recognizes that the identification is hypothetical, but considers it highly probable. Only Tannery, "La Cissoide" pp.43-44 expresses some skepticism, and suggests that the ancient "cissoid" may rather have been an epicycloid or hypocycloid.

Origin of name "cissoid"

It is of some interest in the history of mathematics to answer the question, who was responsible for the identification of Diocles' curve and the name "cissoid". The identification was by no means trivial, since it required informed reading of both Eutocius and Proclus. I regret to say that I am unable to provide the answer. When the name "cissoid" for the curve first appears (in the seventeenth century, to the best of my knowledge), the identification is simply taken for granted. The earliest certain example known to me is in a letter of Roberval written to Fermat in August of 1640: "J'avois fait la même chose en la cissoide" etc. [37]. The context ensures that he is referring to the curve of Diocles. Probably slightly earlier is a reference by Fermat himself in his "Methodus ad disquirendam maximam et minimam": "tangens cissoidis cujus Diocles traditur inventor" [38]. It is highly likely that Descartes is referring to Diocles' curve even earlier, in his *Géométrie* published in 1637: "bienqu'ils ayent aprés examiné la Conchoide, la Cissoide" [39], but since he does not further characterize the curve, the identification is not absolutely secure. Huygens, Newton and others later in the century always call the curve "cissoid" without further justification.

I have searched in vain (but far from exhaustively) in mathematical works of the late sixteenth and early seventeenth centuries for the origin of this nomenclature. It is possible that Fermat himself made the identification: he was certainly well enough read in Greek mathematics to have done so. But if he did, he never made it explicit in any of his published works. The identification could have been made at any time after the publication of Eutocius in 1544 (the first printing of Proclus' commentary took place earlier, as part of the edition of Euclid published by Hervagius at Basel in 1533). But examination of the writings of sixteenth-century authors who are known to have been familiar with both works, such as Commandino and Pierre de la Ramée (or rather such of their writings as I have had access to) has proved fruitless. I must leave the problem to those who are better acquainted with the mathematical literature of that period [40].

---

37 Fermat, *Oeuvres* II p.201.

38 Ibid. I p.159. This was not printed until the publication of Fermat's *Varia Opera* 1679, but according to the editors of *Oeuvres* was sent to Descartes ca. 1638.

39 Descartes, *Géométrie* p.317 (Smith and Latham p.45).

40 The first occurrence of the word (not, however, applied to Diocles' curve) outside Greek is, as far as I know, in the Latin translation of Proclus (1560) by Barozzi (Barocius), p.72: "Cum autem Cissoides, hoc est Haedere similis Linea" etc.

6. Manuscripts and Text of *On Burning Mirrors*

There is a summary description of the manuscript in the Shrine Library, Meshhed, which is the sole basis of the present text, by Golchīn-Maʿānī, *Fihrist* pp. 344-50. The following supplements and occasionally corrects that.

The manuscript is now divided into two parts, numbered 392 (old no. 5593) and 393 (old no. 5521). Golchīn-Maʿānī explicitly states that they were originally a single manuscript [1], and it is obvious from the identity of script and format that this was so. Something of the history of the manuscript can be gathered from the page which was originally the last one in the manuscript, but which is now stuck in as p. 1 of no. 393 (see p.114) [2]. The scribe of the manuscript dates its completion in the year A.H. 867 (= A.D. 1462/3) [3]. In the seventeenth century the manuscript was in the library of the Mughal emperors. At the top of the page is a large seal of Shāh Jahān. In the lower half are a number of certifications in Persian, each accompanied by a seal (?of the imperial librarian), some stating that the manuscript was inspected ("ʿarḍ dīde šod") and all giving a date. The earliest of these is dated in the Ilāhī Era [4], year 94 (= A.D. 1649/50) [5], which indeed falls within the reign of Shāh Jahān. The others are all dated by the Hijra Era: 1087, 10 D̲ū ʾl-Qaʿda (= A.D. 1677, January 14); 1090 (= 1679/80); 1092 (= 1681/2); 1094 (= 1682/3) and finally 1107, 10 Jumādā II (= 1696, January 16) [6]. These all belong to the reign of Aurangzeb. By the

---

1 Golchīn-Maʿānī, *Fihrist* p.350 n.1.

2 This history is summarized by Golchīn-Maʾānī, ibid.

3 "fī taʾrīk̲ sana sabʿa wa-sittīna wa-t̲amānmiʾa". It is possible that a more precise date is indicated in the preceding two lines: "waqaʿa ʾl-farāġ ʿan taḥrīri ʾl-nusk̲ati ʾl-mubāraki bi-ʿawni ʾllāhi taʿālā wa-ḥusni tawfīqihi yawm ? waqt al-ẓuhr", "the release from the writing of the blessed copy occurred, with the help of God on high and the good success he granted, on the day of ? , at the time of midday prayer". The enigmatic word after "yawm" may designate one of the "named" days of the year, but I have failed to find any which resemble it. If one emended it to "al-aḥad" it would mean "Sunday", which is possible even though it would not give a unique date.

4 On this era, used for a short time in the Mughal empire between the reigns of Akbar and Shah Jahān, see Sircar, *Indian Epigraphy* pp.306-07.

5 I dubiously read "5 Shahrivar" ("Shahrivar" also Golchīn-Maʿānī, *Fihrist* p.350 n.1), which would correspond approximately to A.D. 1649, August 21.

6 The last is not accompanied by a seal, but otherwise appears to belong to the same series.

later nineteenth century the manuscript had migrated to Meshhed, for there are two annotations, dated 14 Šawwāl 1270 and Šawwāl 1273 (= 1854 July 10 and 1857 May/June), linking it to the library of the Fāḍiliyya Madrasa in the city. Thence it came to the Shrine Library, with the other books in the Fāḍiliyya Madrasa, in recent years [7]. It is an attractive but unprovable conjecture (suggested to me by W.O. Beeman) that the manuscript came to Meshhed as part of the immense booty which Nādir Shāh brought back from Delhi after his victory over the Mughals in 1739 [8].

The manuscript is all written in the same hand, a careless nastaʿlīq. Diacritical points occur infrequently, and where they do occur are often wrong. Geometrical figures are omitted throughout, but usually a blank space is left where they should have been inserted (see e.g. p. 120). In those parts of the manuscript which I have examined carefully there are many omissions and corruptions (particularly in letters denoting geometrical points). It is obvious that the scribe did not understand much of what he was copying. The manuscript contains at present 154+26 written pages (in nos. 392 and 393 respectively; the written pages in no. 392 are numbered 1-156, but pp. 134-35 are blank). Most pages have 27 lines, the size of the written part being approximately 16.5 x 8.5 cm. [9] It consists now of fourteen mathematical treatises (as detailed below), but must once have contained more, since the original final page (now p. 1 of no. 393) contains the end of a treatise not now in the manuscript. The contents are as follows.

1. No. 392 pp. 1-31. Qusṭā b. Lūqā (ca. 820-ca. 912; Suter no. 77, GAL I², 222-23, S I 365-66), "Book on the reasons for the variations in appearance which occur in mirrors" (Kitāb fī ʿilal mā yaʿriḍu min iḵtilāfi ʾl-manāẓir). This optical work, in 33 chapters, is not in the long list of Qusṭā's writings given by ibn Abī Uṣaybiʿa I pp. 244-45

---

7 Since 1930, when Uktāʾī published a catalogue of the Fāḍiliyya Library (*Fihrist-i kutub-i kitābkhāna-i Madrasa-i Fāḍiliyya;* non vidi). According to Golchīn-Maʿānī, l.c., the manuscript is described (already in two parts) in this catalogue of Uktāʾī. My authority for the transfer of the Fāḍiliyya library to the Shrine Library is Afshar, *Bibliographie* p.28 no.86.

8 See e.g. Datta, *Libraries of India* p.75, for Nādir Shāh's removal of manuscripts from the Imperial Library. Nādir Shāh was a great benefactor of the Shrine Library: see Golchīn-Maʿānī, *Fihrist* index p.527 s.v. "Nādir Shāh Afshār" for references to books he gave to the library. I cannot say whether he made similar benefactions to the Fāḍiliyya Madrasa.

9 Golchīn-Maʿānī, *Fihrist* p.350.

(it is not to be identified with the work "On Burning-Mirrors", Kitāb fī ʾl-marāyā ʾl-muḥriqa), and this copy appears to be unique.

2. pp. 32-35. "Treatise of Didymus on the construction of the mirror by means of which Archimedes burned the ships of the enemy" (Maqāla Dīdīmus fī sanʿati ʾl-mirāʾati ʾllatī aḥraqa bi-hā Aršimīdis marākib al-ʿadū). Though attributed to an author with the good Greek name of Δίδυμος[10], this is surely a pseudepigraphic Islamic work. The author explains that the army of Persia (Īrānšahr) was besieging the town of Archimedes, which was Cos (Qū), by ship from the direction of Samos. It is inconceivable that a Greek of any period could have been so ignorant of the historical facts. The treatise is mathematically absurd, too. It is otherwise extant only in Chester Beatty 5255, 27^v-32^r, which is presumably a direct copy of the Meshhed manuscript, like the treatise of Diocles immediately preceding it (see p. 31).

3. pp. 36-39. Abū ʾl-Futūḥ Aḥmad b. Muḥammad (?)al-Bagnawi (Suter no. 287, GAL S I 857), "On the construction of an equilateral triangle within (another) equilateral triangle such that the ratio between them be any given ratio not less than one to four" (Fī ʿamal muṯallaṯi mutasāwī ʾl-aḍlāʿ fī dāḵil muṯallaṯi mutasāwī ʾl-aḍlāʿ lahu nisba ilayhi mafrūḍa ayyu nisba kānat min al-nisab (al-nisba ms.) allatī laysat aqalla min nisbati ʾl-rābiʿ). The same treatise is in Columbia Or. 45 no. 16 (Awad p. 263) and Leiden Or. 14 pp. 242-45 (*Handlist* p. 431). On the author see also no. 7 below and Kunitzsch, *Der Almagest*, which contains an extensive analysis of his work on the reasons for the errors in the *Almagest* star catalogue.

4. pp. 39-45. "A number of questions of ibn Kišna[11] in refutation of passages in the book 'al-Kāfī' of al-Karajī" (ʿIdda masāʾil li-ibn Kišna fī ʾl-raddi ʿalā mawāḍiʿin min kitābi ʾl-Kāfī lil-Karajī). Consists of a number of extracts from the well-known arithmetical work of al-Karajī (tr. Hochheim, Halle 1878-80; cf. GAS V pp. 328, 403), each followed by the author's remarks on it. The author is the same as in no. 14 below (Suter no. 207). No other copy is known to me.

5. pp. 46-48 (anonymous). "Another way of performing the last proposition of the fifteenth book of the *Elements* (al-šaklu ʾl-āḵir min al-maqālatiʾl-ḵāmisa ʿašr min kitābi ʾl-Uṣūl ʿalā waḍʿ āḵar). The proposition in question, the last of [Euclid] *Elements* XV, is "to inscribe a dodecahedron into a given icosahedron".

---

10 There exists a late Greek metrological treatise under the name of Didymus of Alexandria, printed in Heiberg, *Mathematici Graeci Minores* pp. 4-22.

11 I learn from Russell, *Natural History of Aleppo* I p. 74, that "kishna" is the plant "small vetch" (vicia).

6. pp. 48-81. Ṯābit b. Qurra, "Book on the area of the section of the cone which is called the parabola" (Kitāb fī misāḥa qaṭʿi ʾl-maḵrūṭi ʾllaḏī summiya ʾl-mukāfī). This treatise of the famous ninth-century mathematican also exists in Paris 2457, 25° (GAL I^2 243 no. 14), from which it was translated by Suter, "Ausmessung der Parabel". It is also in Istanbul, Aya Sofya 4832, 3°, 26^v-36^v (Krause p. 455 no. 10). For other mss. see GAS V pp. 269, 402.

7. pp. 81-92. Aḥmad b. Muḥamad b. (?)al-Sūrā, "Treatise explaining what mistake was made by Abū Naṣr al-Fārābī in his commentary on the 17th section of the fifth book of the *Almagest*, with a commentary on that section" (Qawl... fi bayān mā wahima fī-hi Abū Naṣr al-Fārābī ʿinda šarḥihi ʾl-faṣli ʾl-sābiʿ ʿašr mina ʾl-maqālati ʾl-ḵāmisa mina ʾl-majastī wa-šarḥ haḏā ʾl-faṣli). The author is probably the same as no. 3 above (who is given the name "b. al-Sūrā" by ibn Abi Uṣaybiʿa II p. 164). The treatise is not otherwise known, but cf. the similar titles and subjects in the list of Abū ʾl-Futūḥ's works GAL S I 857, nos. 2, 6 and 7. A commentary on the *Almagest* by the famous ninth-century philosopher al-Fārābī is mentioned in Islamic bibliographical works (e.g. ibn al-Qifṭī p. 279, 17-18), but does not appear to be extant. The chapter of the *Almagest* in question concerns the moon's parallax.

8. pp. 92-106. ʿAbd al-Wāḥid b. Muḥammad al-Jūzjānī (11th century; Suter no. 425, GAL S I 828), "Epitome of the arrangement of the spheres" (Ḵulāṣ [sic] tarkībi ʾl-aflāk). The work deals with the order and arrangement of the heavenly spheres, as described by Ptolemy in his *Planetary Hypotheses* (Kitāb al-manšūrāt) and elaborated by Ṯābit b. Qurra, al-Farġānī, ibn Sīnā and others. The same work Leiden Or. 174, 63^v-67^v (*Handlist* p. 148). The author is to be identified with the pupil of ibn Sīnā (so Brockelmann), and hence cannot have lived in the 14th century, (as Suter claims).

9. pp. 106-128. Diocles, "On Burning Mirrors". See p. 31.

10. pp. 128-156. Aḥmad b. Kaṯīr al-Farġānī (9th century, Suter no. 39), a work on the astrolabe. For other manuscripts of work(s) by al-Farġānī on the astrolabe see GAL I^2 250, S I 393. The introduction to one such treatise was translated from ms. Berlin 5790 by Wiedemann, "Einleitungen" (non vidi).

11. No. 393 pp. 2-13. Abū ʾl-Wafāʾ Muḥammad b. Muḥammad b. Yaḥyā al-Būzjānī (the famous 10th-century mathematician and astronomer, Suter no. 167, GAS V 321-25), "On the sum and difference of the sides of squares and cubes" (Fi jamʿ adlāʿi ʾl-murabbaʿāt waʾl-mukaʿʿabāt wa-aḵḏ tafāḍuli-himā). The work is not otherwise known, but its content accords with Abū ʾl-Wafāʾ's interest in algebra, attested by the titles of works attributed to him by the bibliographers

(see Suter and GAS, ll. cc.). It is in answer to a question of Abū Bišr al-Ḥasan b. Sahl the astronomer (? a descendant of the famous astrologer Sahl b. Bišr), and is addressed to the "Shāhānshāh ... al-Mu'ayyid al-Manṣūr" If we identify the latter with the Buwayhid prince Mu'ayyid al-dawla Abū Manṣūr, ruler of Iṣfahān from 976-983, this would add something to our scanty knowledge of Abū 'l-Wafā''s life.

12. pp. 14-17. (anonymous) "Epitome of a statement made by Abū 'l-Rayḥān in his work on the ratios in volume and weight between metals and jewels" (Talḵīṣ kalām ḏakara-hu Abū 'l-Rayḥān fī risāla la-hu fī nisabi 'l-filizzāt wa 'l-jawāhir fī 'l-ḥajm wa 'l-wazn). The work of al-Bīrūnī (fl. 1000) from which this is epitomized is known from al-Bīrūnī's own list of his writings and excerpts in later authors [12], but survives only in photocopies (preserved at the American University of Beirut) of the unique manuscript, Université Saint-Joseph, Bibliothèque Orientale, 223(6), which has now disintegrated.

13. pp. 17-22. Aḥmad b. Muḥammad ʿAbd al-Jalīl al-Sijzī (fl. 1000, Suter no. 185, GAS V 329-34). "Remarks on the nature of the concept of the two lines which Apollonius mentioned in the second book of his *Conics*" (Qawl fī kayfiyya taṣawwur al-ḵattayni 'llaḏayni ḏakara-humā Ablūniyūs al-fāḍil fī 'l-maqālati 'l-ṯāniya min kitābihi fī 'l-maḵrūṭāt). The same treatise is in Leiden Or. 14 pp. 226-31 (*Handlist* p. 180), Istanbul, Reşit 1191 ff. 73-79 (GAS V 332-33) and (probably) Columbia Or. 45 no. 12 (Awad p. 263), i. e. Sezgin's nos. 10, 25 and 28 are all the same work. The "two lines" are the asymptotes of the hyperbola.

14. pp. 22-26. Rašīd al-Dīn Abū Jaʿfar Muḥammad b. Aḥmad b. Muḥammad b. Kišna al-Qummī (Suter no. 207), "Epistle in explanation of the asymptotes" (Risāla fī ibānati 'l-ḵattayni 'llaḏayni yaqrubāni abadan wa-lā yaltaqiyāni). The work is in reply to a question by a certain Abū 'l-Badr ʿAbd al-ʿAzīz b. ʿAlī b. ʿAbd al-ʿAzīz. It is also extant in Chester Beatty 5255 ff. 32-37 (where, like nos. 2 and 10 above, it is presumably a direct copy of this manuscript), in Columbia Or. 45 no. 13 (Awad p. 263) and in Leiden, Or. 14 pp. 232-35 (*Handlist* p. 180). I do not know on what grounds Suter (l. c.), Brockelmann (GAL S I p. 389 no. 7c) and Sezgin (GAS V 336) say that the author was a younger contemporary of al-Sijzī. If correct, this would date him to the 11th century. Brockelmann, on the basis of the Leiden ms., gives his name as b. (?)Kišnab. Cf. no. 4 above.

The treatise of Diocles is also extant in another manuscript, Chester Beatty 5255, 1^v-26^v (see Arberry, *Handlist* pp. 81-82). I

12 See Boilot no.63 pp.196-97.

call this ms. "C". There can be no doubt that it is (at least for the Diocles treatise) a direct copy of the Meshhed manuscript (which I call "M"). The text it presents is very close to that of M, with the same omissions, repetitions and corruptions. More significantly, a number of features in C are explicable as misreadings of M. Thus "min" consistently appears as "fi" in C. A glance at the way it is written in M (e.g. p. 127 line 1, see my p. 136) shows why. The most striking case, however, is p. 107 line 13 of M (§10, see my p. 116), where the scribe repeated in error the words "mitla rubʿi ʾl-kaṭṭī ʾlladī taqwā ʿalayhi ʾl-a" from the previous line, realized his mistake in the middle of the word "ʾl-aʿmida", and crossed the repetition out. However, he crossed it out carelessly, so that his pen skipped the word "ʿalayhi". The copyist of C (who obviously understood not a word of what he was copying), slavishly omitted the rest of the repeated phrase, but copied the nonsensical "ʿalayhi" (C, 2^r line 10). No further argument is necessary to prove C's direct dependence on M, though many could be adduced. I have no doubt that the second and third items in C are also directly copied from M (see pp. 28 and 30, items 2 and 14) but I have not made a detailed comparison of those works. Thus C can be eliminated as a witness for Diocles. Accordingly, I refer throughout only to the readings of M.

M was written very carelessly (see the general description, p. 27), and mathematical grounds alone force one to assume numerous corruptions (fortunately, most of a trivial kind) in the Diocles treatise. I have not hesitated to emend the text where it seemed necessary. I am confident that I have restored the sense, if not the exact wording, of the original in passages of mathematical demonstration. The same is true, *mutatis mutandis*, for my reconstructions of the missing diagrams. I am much less confident of my text in certain expository passages, notably §§ 16-37.

Besides these merely mechanical scribal errors, the text has suffered severely from deliberate interpolation. I believe that Props. 6, 9 and 14 cannot be attributed to Diocles (for my reasons see the commentary on those propositions, pp.161-2,168-9,173). There are also some individual sentences which are highly suspicious (see notes on §§ 53, 154). It seems highly probable, though unprovable, that these interpolations occurred after the treatise was translated into Arabic [13]. Nevertheless, I have not eliminated these spurious passages from the text, preferring to present the Arabic version of Diocles "tel quel".

---

13 Arguments to support this can be drawn from vocabulary. See note on §53 (use of "watar"). The word "ṣūra" for "diagram" occurs only in a section considered spurious on other grounds (§§232, 235).

Editorial Procedures

All numerical references in the critical apparatus and (unless otherwise specified) in the commentary are to the numbers of the sections into which I have arbitrarily divided the text. The end of each section is marked by the sign ı in both text and translation. In the text my supplements are enclosed in angled brackets, while passages which I have deleted as scribal additions or repetitions are enclosed in square brackets. All other changes from the manuscript, except changes in the pointing, are reported in the apparatus, in the following way. I give first,. as catchword(s), the correct reading, as it stands in the text above, then, separated by a colon, the manuscript reading. Different items within the same section are separated by a semi-colon. If the relevant word or phrase occurs more than once within the same section, a raised 1, 2, etc. after the catchword(s) indicates that the reference is to the first, second, etc. occurrence. Only the readings of ms. M are reported, since it is the only independent witness to the text (see Introduction pp. 30-31). Since it is so carelessly written, I do not report the numerous cases where I supplement or even change the pointing of letters. Nor, in general, do I mention or discuss alternative ways of reading the traces in the manuscript, though there are many places which, from a palaeographical point of view, could be read differently. The photographs of the manuscript, pp. 114-137, afford a means of checking my text and apparatus.

The text figures are entirely my reconstruction, since all are omitted in the manuscript. Only for Figs. 7, 8, 10, 12 and 13 do we have the version of Eutocius for comparison (cf. Figs. VIII-XII in Appendix A), and I have indeed based my reconstruction of those figures partly on those in the mss. of Eutocius, though it is likely that in the Arabic text the figures were the mirror images of those in the Greek, as is often (but by no means always) the case. I have represented the lettering of the figures by a consistent system, which corresponds to the lettering of the Arabic as detailed below. From the Arabic we can in turn reconstruct the Greek lettering, since Arabic mathematical texts translated from the Greek represent the letters by the Arabic letters in the numerical ("abjad") order. On this point see Gandz, "Der Hultsch-Cantorsche Beweis", for a good summary of the older literature, though the article needs correcting on several points of fact, and Gandz's contention that the Arabic order is independent of the Greek is quite unacceptable. Each Greek letter is represented by the Arabic letter which has the same numerical value. There are no general exceptions to this rule, and very few individual exceptions, and there can be no doubt that the translators were fully aware of the correspondence, as can be amply demonstrated e.g. from the Arabic translation of Apollonius' *Conics*. The correspondences in our text are as follows.

My translation		Arabic		Greek
A	=	alif	=	Α
B	=	bā	=	Β
G	=	jīm	=	Γ
D	=	dāl	=	Δ
E	=	hā	=	Ε
Z	=	zāy	=	Ζ
H	=	ḥā	=	Η
Θ	=	ṭā	=	Θ
K	=	kāf	=	Κ
L	=	lam	=	Λ
M	=	mīm	=	Μ
N	=	nūn	=	Ν
S	=	sīn	=	Ξ
O	=	ʿayn	=	Ο
P	=	fā	=	Π
Ψ	=	ṣād	=	Ϙ
Q	=	qāf	=	Ρ
R	=	rā	=	Σ
C	=	šīn	=	Τ
T	=	tā	=	Υ
F	=	ṯā	=	Φ
X	=	ḵā	=	Χ
U	=	ḏāl	=	Ψ
Φ	=	ḍād	=	Ω

wāw (Greek Ϛ), yā (Greek I) and ẓā (Greek Ϡ) do not occur in our text. This is not surprising, since the corresponding Greek letters very rarely occur in geometrical diagrams.

On Burning Mirrors

1 In the name of God, the merciful, the compassionate.
O God, grant long life.|

2 The book of Diocles on burning mirrors.|

3 He said: Pythion the Thasian geometer wrote a letter to Conon
in which he asked him how to find a mirror surface such that
when it is placed facing the sun the rays reflected from it
4 meet the circumference of a circle.| And when Zenodorus
the astronomer came down to Arcadia and was introduced
to us, he asked us how to find a mirror surface such that
when it is placed facing the sun the rays reflected from it
5 meet a point and thus cause burning.| So we want to explain
the answer to the problem posed by Pythion and to that posed
by Zenodorus; in the course of this we shall make use of
6 the premisses established by our predecessors.| One of
those two problems, namely the one requiring the construc-
tion of a mirror which makes all the rays meet in one
point, is the one which was solved practically by Dositheus.|
7 The other problem, since it was only theoretical, and
there was no argument worthy to serve as proof in its
case, was not solved practically. We have set out a
compilation of the proofs of both these problems and
elucidated them.|

8 The burning-mirror surface submitted to you is the
surface bounding the figure produced by a section of a

1 بسم الله الرحمن الرحيم اللّهمّ اعمر.

2 كتاب ذيوقليس في المرايا المحرقة.

3 قال انّ فوثيون المهندس الذي من اهل تاسيس
كتب الى قونون رسالة سأله فيها كيف نجد بسيط مرآة حتى
وضع قبالة الشمس اجتمعت الشعاعات التي تنعطف منه الى
4 خطّ محيط بدائرة. واما زينوذروس المنجّم فانّه لمّا طرأ الى
ارقاذيا وقدّم لنا سألنا كيف نجد بسيط مرآة حتى وضع قبالة
الشمس اجتمعت الشعاعات التي تنعطف منه الى نقطة
5 فاحرقت. فاما نحن فانّا نروم ان نبيّن الجواب فيما سأل عنه
فوثيون وما سأل عنه زينوذروس ونستعمل في ذلك المقدّمات
6 التي قدّمها مَن كان قبلنا. واحدى هاتين المسئلتين وهي
التي يطلب فيها عمل مرآة يجتمع شعاعها الى نقطة واحدة
7 كانت التي عملها ذوسيثاوس. واما المسئلة الاخرى فانّها لمّا
كانت علماً فقط ولم يكن لها قول يستحقّ ان يشهد به
لم تعمل. وقد بيّنّا تأليف براهين كلّ واحدة من هاتين
8 المسئلتين واوضحناها. وبسيط المرآة المحرقة الذي رفع اليك هو

---

3 المهندس: المهد (supra in rasura س)

4 زينوذروس: ابيوذامس؛ طرأ: بطر؛ لنا: فمها

5 سأل[1]: ساد؛ زينوذروس: اينوذامس

6 كانت التي: كان الدي

right-angled cone (i. e. parabola) being revolved about the
9 line bisecting it (i. e. its axis).| It is a property of that
surface that all the rays are reflected to a single point,
namely the point (on the axis) whose distance from the
surface is equal to a quarter of the line which is the para-
meter of the squares on the perpendiculars drawn to the axis
10 (i.e. the ordinates).| Whenever one increases that surface by
a given amount, there will be a (corresponding) increase in
11 the above-mentioned conic section.| So the rays reflected
from that additional (surface) will also be reflected to
exactly the same point, and thus they will increase the
12 intensity of the heat around that point.| The intensity of
the burning in this case is greater than that generated from
a spherical surface, for from a spherical surface the rays
are reflected to a straight line, not to a point, although
people used to guess that they are reflected to the center;|
13 the rays which meet at one place in that (i. e. a spherical)
surface are reflected from the surface (consisting) of a
14 spherical segment less than half the sphere,| and (even)
if the mirror consists of half the sphere or more than
half, only those rays reflected from less than half the
sphere are reflected to that place.|

15 The problem posed by Pythion is also solved by a
section of a right-angled cone being revolved with another

البسيط الذي يحيط بالشكل الذي يحدث من قطع المخروط
9 القائم الزاوية اذا ادير حول الخطّ الذي يقسمه بنصفين. فانّه
يعرض لهذا البسيط ان ينعطف جميع شعاعاته الى نقطة واحدة
وهي النقطة التي بعدها من ذلك البسيط مثل ربع الخطّ الذي
10 تقوى عليه الاعمدة التي تخرج الى السهم. وكلّما زيد في ذلك
البسيط زيادة معلومة [على قطعة دائرة] يزاد في قطع المخروط الذي
11 ذكرنا. فانّ الشعاعات التي تنعطف من تلك الزيادة تنعطف
ايضًا الى تلك النقطة بعينها فتزيد في قوّة الحرارة التي حول
12 النقطة. وقوّة هذا الاحراق اقوى من قوّة الاحراق الذي يكون عن
بسيط الكرة وذلك انّ انعطاف الشعاعات من بسيط الكرة انّما
يصير الى خطّ مستقيم لا الى نقطة وان كان قد ظنّ قوم
13 انّها تصير الى المركز. والشعاعات التي تجتمع من هذا
البسيط الى موضع واحد انّما تنعطف من بسيط قطعة كرة اقلّ
14 من نصف الكرة. وان كانت المرآة من نصف كرة او من
اكثر من نصفها لم ينعطف الى ذلك الموضع غير تلك
15 الشعاعات التي انعطفت من اقلّ من نصف الكرة. والمسئلة

---

9 التي[1]: نقسمه سصفين فانه نعرض لهذا السيط ان سعطف حمىع add. sed del.

10 بسيط: مثل ربع الحط الذي نقوى علىه الا add. sed del.

13 هذا: هذه

16 kind of revolution, and we shall explain that later.| (Thus)
an ingenious method has been found for a burning-mirror
to burn without being turned to face the sun; instead it
is fixed in one and the same position, and indicates the
17 hours of the day without a gnomon.| It does this by burning
a trace to which the rays are reflected: the reflecting
produces a trace for the position of the hour which is
sought. This statement is amazing, namely that there is
no need to turn the mirror, but that (what we have
described) results merely from the above-mentioned
figure.|

18 We discuss first of all an assumption constantly made
by the astronomers, namely that every point on the earth
19 can be treated as the center of the earth.| Sometimes
people who try to discredit the mathematical scientists
and say that they construct their subject on a weak
20 foundation scoff (at this):| for some of them (the scientists)
assert that the (size of the) radii of the spheres is known
and that each one is greater than the one (next to it) by
more than 30 million stades, while others assert that it
21 (is greater by) more than 50 million stades.| People
were inclined more to this second opinion, because they
trusted the doctrines of the ancients in this matter;
but they say that if the way can be found to avoid using

التي سأل عنها فوثيون تعمل ايضًا بقطع المخروط القائم الزاوية
16 اذا ادير ضربًا ما من الادارة وسنبيّن ذلك فيما بعد. وقد
احتيل في عمل مرآة محرقة تحرق من غير ان تقلب بازاء
الشمس وهي ثابتة في موضع واحد بعينه وتبيّن ساعات
17 النهار من غير مقياس. وذلك انّها تحرق اثرًا ما تنعطف
اليه الشعاعات وانعطافها يكون اثرًا الى موضع الساعة المطلوبة.
وهذا القول شيء يتعجّب منه وذلك انه لا يحتاج فيه الى
[شيء] تقليب المرآة لكنّه يحدث عن الشكل الذي ذكرنا
18 فقط. ونحن ذاكرون اوّلاً مصادرة يستعملها المنجّمون وهي انّ
كلّ نقطة [يستعمل] من النقط التي على الارض فهي تقوم
19 مقام مركز الارض. وقد يضحك قوم ويشنّعون على اصحاب
العلوم التعليميّة ويقولون انّهم يبتنون امرهم على اساس
20 ضعيف. ويزعم بعضهم انّه قد علم انصاف اقطار الاكر
[وانّه واحد بعضها] وانّ الواحد منها اكبر من الآخر باكثر
من ثلثين الف 〈الف〉 سطاذيون ويزعم بعضهم انّ ذلك
21 اكثر من خمسين الف 〈الف〉 سطاذيون. وميل الناس الى
هذا الرأي الثاني اكثر لانّهم وثقوا فيه باقاويل القدماء. ويقولون
انّه ان وجد المبيل الى ان لا يستعمل فيها هذا الاصل

---

18 النقط: النعطه

20 ويزعم²: وزعم

21 القدماء: القد

this principle, and we are not forced to use it by the
requirements of the subject of time-measuring intruments
22 which use the shadow, then it is best not to use it.| We
shall proceed with our discussion in order to explain
what they were doubtful about. We say, then, that the
above-mentioned point (i. e. any point on the earth's surface) can be treated as the center of the earth and of the
23 universe.| In this we must state the cogent analogy which
we use in this matter and in others, and we state that
the phenomena occuring in the case of the gnomons are
similar to what would occur if they really lay at the center
24 (of the earth).| But those time-measuring instruments
using the shadow which indicate the hours without having
a gnomon in them reach a degree of minute accuracy
such as cannot be attained in this matter by any other
25 kind (of instruments).| It may also be possible for you
to examine the way these instruments are used, if you
would care to do so, because (firstly) there is in them
something to evoke your astonishment: for we are able to
do something which others have (merely) talked about in
connection with this subject, and (to decide) whether they
26 hit the mark or not in what they wrote about it;| and
because, secondly, to sum up, none of the elements
needed in the above-mentioned time-measuring instru-
27 ments is lacking in our operation.| I believe that the
operation we must expound to you is something you may

ولم يضطرّنا اليه الحاجة في امر آلات الساعات التي يستعمل
22 فيها الظلّ فالاصلح ان لا يستعمل. ونحن نتجاوز هذا
الموضع حتى نبيّن ما شكّوا فيه فنقول انّ النقطة التي ذكرنا
23 تقوم مقام مركز الارض والعالم، وينبغي ⟨ان نبيّن⟩ في
ذلك القياس الواجب الذي نستعمله في هذا الموضع وغيره
ونبيّن انّ الذي يعرض في امر المقاييس شبيه بما كان يعرض
24 فيها لو وضعت على الحقيقة في المركز. فانّ آلات الساعات
التي يستعمل فيها الظلّ ما كان منها دالّاً على الساعات
من غير مقياس يكون فيه يبلغ في الاستقصاء والحقيقة حتى لا
25 يكون في ذلك دون غيره. وقد يمكنك [ان] ان تنظر ايضاً في
عمل هذه الآلات ان احببت ذلك لانّه فيها شيء يستحقّ ان
تتعجّب منه وذلك انّه يمكنّا ان نعمل ما قاله في ذلك
26 غيرنا وهل اصابوا فيما كتبوا من ذلك او لم يصيبوا، ولانّه
مع ذلك بالجملة لا يذهب ⟨علينا⟩ في هذا العمل شيء نحتاج
27 اليه في آلات الساعات التي ذكرنا. وانا ارى انّ الذي
ينبغي ان نبيّن عمله لك هو ⟨ما⟩ يعلم بسهولة وايجاز [انه]

---

22 شكّوا: نتكوا

23 المقاييس: العماس

24 الساعات: الشعاعات (bis)

25 لانّه: لان

27 نبيّن عمله لك: سه للحلوك؛ يعلم: علموا

understand easily and briefly, so it does not require that
28 you give it close attention and study.| As for the matter
of the gnomons used by the astronomers, they achieve
great accuracy when they are made according to the old
methods which used to be employed in making time-
measuring instruments in which the shadow is used.|
29 But many of the surfaces in which it (the shadow) is used
are impossible to make, and many of them are very
30 difficult to make.| In short, you must realize that the
knowledge of this is difficult and requires care and
perseverance; whoever has spent pains on it will attest
the truth of what we say.|

31 Perhaps you would like to make two examples of a
burning-mirror, each having a diameter of two cubits,|
32 one constructed on the circumference of a circle (i. e.
spherical), the other on a section of a right-angled cone
(i.e. parabolic), so that it may be possible for you to measure
the burning-power of each of them by the degree of its
33 efficiency.| So one knows the base of their burning-powers,
and (then) measuring the (relation between) the burning
(-power) of one and that of the other is a matter requiring
34 observation:| that is to say, if the mirror-surface with
a diameter of the amount of one foot burns the whole of the
burning-area which heats up in (a piece of) wood, then it
is more likely to burn (it) easily when its diameter is

28 فما يستحقّ ان تنظر فيه وتتعلّم. فاما امر المقاييس التي
يستعملها المنجّمون فانّها مستقصاة صحيحة اذا عملت على الرسوم
القديمة التي كان يعمل عليها آلات الساعات التي يستعمل فيها
29 الظلّ. ولكنّ كثيرًا من البسط التي يستعمل فيها لا يمكن
30 ان تعمل وكثير منها يعسر عملها جدًّا. وينبغي ان تعلم
بالجملة انّ معرفة ذلك عسرة ⟨و⟩تحتاج الى عناية ومواظبة ومَن
31 كان قد عانى ذلك فهو يشهد بصدق ما قلنا. وقد تحبّ ان
تعمل مرآتين من المرايا المحرقة يكون قطر كلّ واحدة منهما
32 ذراعين، فتكون احديهما معمولة على خطّ محيط بدائرة والاخرى
على قطع المخروط القائم الزاوية حتى يمكنك ان تقيس قوّة
33 كلّ واحدة منهما في الاحراق بقوّة صلاحيتها. فيعلم اصل
قويهما احراقًا وقياس احراق واحدة منهما الى احراق الاخرى
34 امر يستحقّ ان يشاهد، وذلك انّ سطح المرآة التي قطرها
مقدار قدم واحدة ان كانت تحرق جميع ما يدفؤ من موضع

---

29 البسط التي: السيط الذي

30 ومن: كل دلك add. sed del.؛ عانى: عانا

31 قطر: نطر

32 احديهما: احداهما

33 قويهما: اقواهما؛ منهما: منها

34 يدفؤ: يدفوا

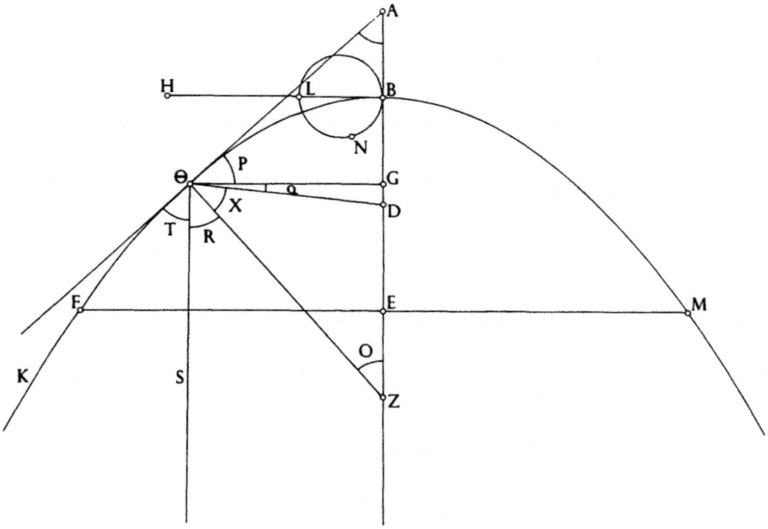

Fig. 1

35 seven times that amount.| For when the burning-power is multiplied by seven, the difference between it (such a mirror) and the original mirror must be very great.|

36 We believe that it is possible to make a burning-instrument of glass such that it has a special property, namely that one can make lamps from it which produce fire in temples and at sacrifices and immolations, so that the fire is clearly seen to burn the sacrificial victims;|
37 this occurs, as we are informed, in certain remote cities, especially on the days of great celebration: this causes the people of those cities to marvel. That is something which we too shall do.|

38 Let there be a parabola KBM, with axis AZ, and let Prop.
half the parameter of the squares on the ordinates be line 1
39 BH.| Let BE on the axis be equal to BH, and let BE be bisected at point D. Let us draw a line tangent to the section at an arbitrary point, namely line ΘA, and draw
40 line ΘG as ordinate to AZ.| Then we know that

الاحراق في الخشب فهي أَحرى ان تحرق بسهولة اذا كان
35 القطر سبعة امثال ذلك، لانّه اذا ضوعفت قوّة الاحراق سبع
مرّات لم يمكن ألّا يكون الفصل فيما بينها وبين الاولى
36 فصلاً كبيرًا قويًّا. ونحن نرى انّه يمكن ان تعمل آلة محرقة
من زجاج حتى يعرض فيها شيء خاصّ وهو انّه تتّخذ منها
مصابيح فوجد منها نار في الهياكل وعند الذبائح والقرابين
37 حتى يرى انّ هذه النار تحرق الذبائح، كما بلغنا انّه يكون
في بعض المدن النائية عنّا وخاصةً في الايّام التي يعظمها فيعجب
اهل تلك المدن من ذلك. وهذا شيء نفعله نحن ايضًا.
38 ليكن قطع المخروط القائم الزاوية كبم والخطّ الذي ⟨أ⟩
يقسمه بنصفين خطّ از ونصف الخطّ الذي تقوى عليه الاعمدة
39 خطّ بح، وليكن خطّ به و⟨هو على⟩ الخطّ الذي يقسمه
مساويًا لخطّ بح وليقسم خطّ به بنصفين على نقطة د.
ولنخرج خطًّا يماسّ القطع كيف ما وقع وهو خطّ طا
40 ولنخرج خطّ طج عمودًا على از. فقد علمنا انّ خطّ اب مساو
لخطّ ⟨بج⟩ وانّ الخطّ الذي يخرج من نقطة ط من خطّ طا
41 على زوايا قائمة يلقى خطّ از خارج نقطة ه. فلنخرج خطّ زط

---

36 في: من
39 به: ه؛ خطّ طج: خطا طج
40 يلقى: يلقا

AB = BG

and that the line drawn from θ perpendicular to θA meets
41 AZ beyond E.| So let us draw Zθ perpendicular to θA, and
join θD.

Then GZ = BH
and HB = BE,
so GZ = BE.|

42 We subtract GE, common (to GZ and BE), then the remainder

GB = EZ.
But GB = BA,
so AB = EZ.|
43 And BD = DE, because BE is bisected at D,
so the sum AD = DZ.|

44 And because triangle AθZ is right-angled and its base AZ is bisected at D,

AD = Dθ = DZ.
So $\hat{O} = \hat{X}$
and $\hat{A} = \widehat{PQ}$.|

45 So let a line parallel to AZ pass through θ, namely line θS.

Then $\hat{O} = \hat{R}$, which is alternate to it,
and $\hat{O} = \hat{X}$,
so $\hat{X} = \hat{R}$ also.|
46 And $\widehat{PQ}X = \widehat{RT}$, right angles,
so $\hat{T} = \widehat{PQ}$, remainders.|

47 So when line Sθ meets line Aθ it is reflected to point D,
forming equal angles, PQ and T, between itself and the
48 tangent Aθ.| Hence it has been shown that if one draws from
any point on KBM a line tangent to the section, and draws

من خطّ طا على زوايا قائمة وليوصل خطّ طد. ⟨ف⟩يكون جز
مساويًا لخطّ بح وخطّ حب مساو لخطّ به فخطّ جز
42 مساو لخطّ به. ونسقط خطّ جه المشترك فيبقى خطّ جب
مساويًا لخطّ هز. وخطّ جب مساو لخطّ با فخطّ اب مساو
43 لخطّ هز. وخطّ بد مساو لخطّ ده وذلك انّ خطّ به قسم
44 بنصفين على نقطة د فجميع خطّ اد مساو لخطّ دز. ولانّ
مثلّث اطز قائم الزاوية وقد قسمت قاعدته از بنصفين على
نقطة د صارت خطوط اد دط دز متساوية. فزاوية ع مساوية
45 لزاوية خ وزاوية ا مساوية لزاوية فق. فليجاز على نقطة ط
خطّ مواز لخطّ از وهو خطّ طس. فزاوية ع مساوية لزاوية ر
المبادلة لها وزاوية ع مساوية لزاوية خ فزاوية خ ايضًا مساوية
46 لزاوية ر. وزاوية فقخ القائمة مساوية لزاوية رت القائمة فزاوية
47 ت الباقية مساوية لزاوية فق الباقية. فخطّ سط اذا لقي خطّ
اط انعطف الى نقطة د فاحدث فيما بينه وبين الخطّ
48 ⟨الماسّ⟩ اط زاويتي فق [ر] ت المتساويتين. فقد تبيّن
انّه ان اخرج من نقطة ما من النقط التي على خطّ كبم

---

41 مساو: مساويا (bis)

42 ونسقط: وسقط

44 دز: دل

46 رت: ا ت

the line connecting the point of tangency with point D, e.g.
49 line θD, and draws line Sθ parallel to AZ,| then in that case
line Sθ is reflected to point D, i.e. the line passing through
point θ is reflected at equal angles from the tangent to the
50 section.| And all parallel lines from all points on KBM have
the same property, so, since they make equal angles
with the tangents, they go to point D.|

51 Hence, if AZ is kept stationary, and KBM revolved
(about it) until it returns to its original position, and a
concave surface of brass is constructed on the surface
described by KBM, and placed facing the sun, so that
52 the sun's rays meet the concave surface,| they will be re-
flected to point D, since they are parallel to each other.
And the more the (reflecting) surface is increased, the
greater will be the number of rays reflected to point D.|

53 But if we cut off line BθF, making it equal to line
BM, and join FM, which is like a chord of section FBM,
and set up on AZ a plane perpendicular to the established
plane (i.e. the plane of paper), so that FM is perpen-
54 dicular to that plane,| and, keeping FM stationary, re-
volve section FBM (about FM) until it returns to its
55 original position,| and (then) construct a surface of brass
conformal to the resulting concave survace, and place
it facing the sun, the rays will be reflected from its whole

خطّ مماسّ للقطع ووصل الخطّ الذي بين نقطة الماسّة ونقطة
49 د مثل خطّ ط د واخرج خطّ س ط مواز لخطّ ا ز، فانّ خطّ
س ط اذًا انعطف الى نقطة د فالخطّ الذي يمرّ بنقطة ط
‹وينعطف من الخطّ› الماسّ للقطع يكون انعطافه على زوايا
50 متساوية، وجميع الخطوط المتوازية التي تخرج من جميع النقط
التي على خطّ ك ب م يكون حالها هذه الحال فاذا احدثت مع
51 الخطوط الماسّة زوايا متساوية سارت الى نقطة د. فان اثبت
خطّ ا ز وادير خطّ ك ب م حتى يعود الى الموضع الذي منه
ابتداءً وعمل سطح مقعّر من صفر على السطح الذي يمرّ فيه
خطّ ك ب م ووضع قبالة الشمس فلقيت شعاعات الشمس
52 البسيط المقعّر، انعطفت الى نقطة د لانّها متوازية وكلّما كان
البسيط اعظم كانت الشعاعات التي تنعطف الى نقطة د
53 اكثر. وان فصلنا خطّ ب ط ث وجعلنا‹ه› مساويًا لخطّ ب م
ووصلنا خطّ ث م الذي كوتر قطع ث ب م واقمنا على خطّ
ا ز سطحًا قائمًا على السطح الموضوع وكان خطّ ث م قائمًا
54 على ذلك السطح على زوايا قائمة، واثبتنا خطّ ث م ثمّ
ادرنا قطع ث ب م حتى يعود الى الموضع الذي منه ابتداءً،
55 وعملنا سطحًا من صفر شبهًا بالسطح المقعّر الحادث
ووضعناه قبالة الشمس انعطفت الشعاعات من جميع بسيطه الى

---

53 وان: بان؛ كوتر: بوتر؛ قطع: قطعه

area to the circumference of a circle (lying) in the plane
56 on AZ.| The reason is that since point D is revolved to-
gether with section FBM, the rays will be reflected to
the circumference of the circle produced by the revolution
of this point. The radius of this circle comes out equal
57 to line DE.| So if we want the circle to be any desired
size, we make DE equal to (that circle's) radius, and
draw BE perpendicular to FM, and perform those other
58 operations exactly (as we did before).| You should realize
that one must not use the whole of the surface we des-
cribed, because otherwise MBF would come in line with
59 the sun and thus obstruct it.| But if we use half of the above-
mentioned surface, which is what results when the surface
is cut by some plane passing through FE, we get the
desired effect. But (in that case) burning does not occur
on the whole of the circumference of the above-mentioned
circle, but (only) on half of it.|

60 But if we erect a plane through BH so that AB is per-
pendicular to it and it is perpendicular to the plane through
FBM, and draw in it some circle, BLN, such that point B
61 lies on the circumference of BLN,| and make segment BM
of the section revolve about it until it returns to its (original)
position, in such a way that line BZ remains perpendicular

56 خطّ محيط بدائرة في السطح القائم على آز، وذلك انّ نقطة د
اذا اديرت مع قطع تبم انعطفت الشعاعات الى الخطّ
المحيط بالدائرة التي تحدث عن دوران هذه النقطة. ويصير الخطّ
الذي يخرج من المركز الى الخطّ المحيط بتلك الدائرة مساويًا
57 لخطّ ده. فان اردنا ان تكون الدائرة بايّ مقدار اردنا جعلنا
خطّ ده مساويًا للخطّ الذي يخرج من مركزها الى الخطّ
المحيط بها واخرجنا خطّ به على خطّ تم على زوايا قائمة
58 وعملنا تلك الاشياء الباقية باعيانها. وينبغي ان تعلم انّه لا
ينبغي ان يستعمل البسيط الذي ذكرنا كلّه حتى لا يحاذي
59 خطّ مبت الشمس فيكون ساترًا لها. وان استعملنا نصف
السطح الذي ذكرنا كأنّ الذي يحدث ⟨اذا⟩ يقطع البسيط
سطح ما يمرّ بخطّ ته كان ما اردنا. ولم يقع الاحراق
على جميع الخطّ المحيط بالدائرة التي ذكرنا بل على نصفه.
60 فان اقمنا سطحًا يمرّ بخطّ بح ويكون خطّ اب قائمًا عليه
على زوايا قائمة ويكون قائمًا على السطح الذي يمرّ بخطّ
تبم على زوايا قائمة ورسمنا فيه دائرة ما وهي دائرة بلن
61 حتى تكون نقطة ب على الخطّ المحيط بدائرة بلن، وادير عليه

---

58 مبت: به بت

59 نصفه: صعه

60 السطح: سطح

to the above-mentioned plane (the plane through BH) through-
62 out the revolution of the segment,| the result of that will be
a certain surface, and point D will describe a certain
circle to whose circumference the rays will be reflected.
The perpendicular distance of that surface from the circle
to whose circumference we want the rays to be reflected
63 will be equal to BD.| And not only it is possible for re-
flection to occur to the circumference of a circle, but
it might also be possible for it to occur to any other curve
64 we wish,| as follows: line BH has been drawn from point
65 B of section KBM at right angles to AB.| Let there be
(drawn) in the plane through BH some curve, BLN, of the
same size and shape as the curve to which we want re-
66 flection of the rays to occur.| Let segment BM of the
section, which is tangent to the curve at point B of curve
BLN, be revolved around curve BLN so that its revolution
is at right angles to the established plane (i. e. the plane of
the paper), and so that line ZB in it (segment BM), which
is tangent to curve BLN, remains fixed in its original
67 position.| It is evident that point D produces in the plane
parallel to the plane through BH a curve equal and
similar to BLN.

قطعة بم من القطع ⟨حتى تعود⟩ على حالها وكان خطّ بز
قائمًا على السطح الذي ذكرنا على زوايا قائمة في جميع دوران
62 القطعة، حدث من ذلك بسيط ما ورسمت نقطة د دائرة ما
تنعطف الشعاعات الى الخطّ المحيط بها ويصير بعد عمود ذلك
البسيط من الدائرة التي نريد ان تنعطف الشعاعات الى الخطّ
63 المحيط بها مثل خطّ بد. وليس انّما يمكن ان يكون
الانعطاف الى خطّ محيط بدائرة فقط بل قد يمكن ان يكون
64 ايضًا الى خطّ آخر ايّ خطّ اردنا، وذلك [وذلك] انّه قد
اخرج من نقطة ب من قطع كبم خطّ بح ⟨على خطّ
65 اب⟩ على زوايا قائمة. فليكن في السطح الذي يمرّ بخطّ
بح خطّ ما وهو خطّ بلن بالمقدار الذي نريد ان يكون
66 الخطّ الذي تنعطف اليه الشعاعات. وليدار على خطّ بلن
قطعة بم من القطع الماسّة للخطّ على نقطة ب من خطّ
بلن وليكن دورانها على زوايا قائمة على السطح الموضوع
[الذي هو السطح الذي يمرّ بخطّ بح] وليكن فيه خطّ زب
67 الماسّ لخطّ بلن لازبًا بوضعه الاوّل. فهو بيّن انّ نقطة
د تحدث في السطح الموازي للسطح الذي يمرّ بخطّ بح
خطًّا مساويًا لخطّ [بح خطًّا مساويًا] بلن شبيهًا به ٥

---

63 يمكن[1]: يكون

66 وليدار: ولدار؛ بوضعه: لوضعه

67 بلن: لبذ؛ ٥: و

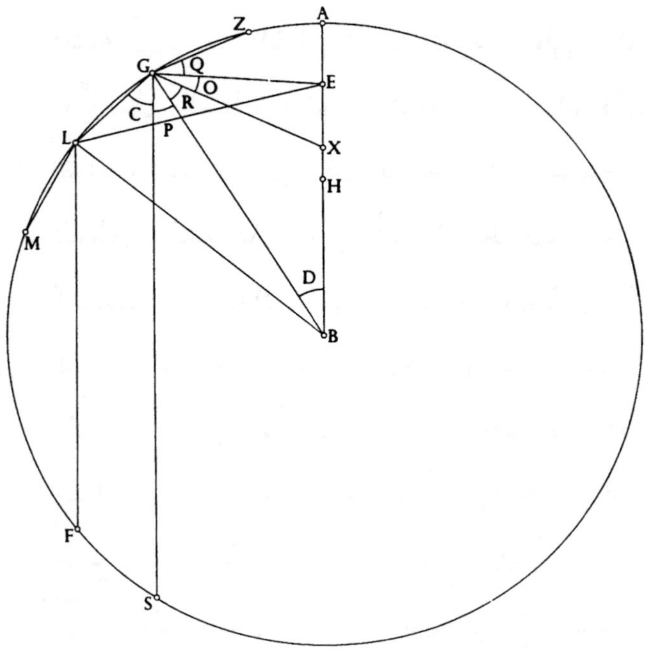

Fig. 2

68 Let FAM be the circumference of a circle whose center is B, and let the circumference be met by some line, AB. Let there be drawn from any two points on it (the circumference) two lines parallel to AB, namely LF, Prop. 2
69 GS.| Join BL, BG, and cut off arcs LM, GZ equal to arc LG. Then

$\widehat{BLM} = \widehat{BLG}$.|

70 And if we make $\widehat{ZGX} = \widehat{LGS}$

and $\widehat{GLE} = \widehat{FLM}$,

then $\widehat{ELB} = \widehat{BLF}$, remainders (of $\widehat{BLG} - \widehat{GLE}$ and $\widehat{BLM} - \widehat{FLM}$).|

71 But $\widehat{FLB} = \widehat{LBE}$, alternate,

so $\widehat{LBE} = \widehat{ELB}$,

so BE = LE.|

68 ليكن خطّ تام محيط بدائرة مركزها نقطة ب وليلق 〈ب
الخطّ المحيط بها خطّ ما وهو خطّ اب وليخرج من نقطتين
فيه ايّ نقطتين كانت خطّان موازيان لخطّ اب وهما خطّا
69 لت جس، وليوصل خطّا بل بج 〈وليفصل قوسا لم
جز مساويتان لقوس لج〉. فتكون زاوية بلم مساوية
70 لزاوية بلج. واذا جعلنا زاوية 〈زجخ مساوية لزاوية
لجس وزاوية〉 جله مساوية لزاوية تلم كانت زاوية
71 هلب الباقية مساوية لزاوية بلت الباقية. ولكنّ زاوية
تلب مساوية لزاوية لبه المبادلة لها فزاوية لبه مساوية
72 لزاوية هلب فخطّ به مساو لخطّ له. ولكنّ خطّ له
اطول من خطّ ها وذلك انّ خطّ ها اقصر الخطوط التي تخرج
73 من نقطة ه الى الخطّ المحيط بالدائرة. فخطّ به اطول من خطّ
ها وهو ايضًا اطول من خطّ هج لانّ خطّ له اطول من
74 خطّ هج. فزاوية عر اعظم من زاوية د والزاوية المبادلة لزاوية
د هي زاوية ف فزاوية عر اعظم من زاوية ف. وجميع زاوية
شف مساوية لزاوية رعق فزاوية ش اعظم من زاوية ق.
75 ولكنّ زاوية قع مساوية لزاوية ش فالزاوية الباقية التي هي
زاوية ر مساوية لزاوية ف الباقية. وزاوية ف مساوية لزاوية د

---

68 وليلق: وللقي

69 خطّا: حط ا

75 زاوية ر: ب ر

72 But LE > EA, because EA is the shortest of (all) the lines drawn from E to the circumference of the circle.|

73 So $BE > EA$

and also $BE > EG$, since $LE > EG$.|

74 So $\widehat{OR} > \hat{D}$;

and the angle alternate to $\hat{D}$ is $\hat{P}$,

so $\widehat{OR} > \hat{P}$.

But the whole of $\widehat{CP} = \widehat{ROQ}$,

so $\hat{C} > \hat{Q}$.|

75 But $\widehat{QO} = \hat{C}$,

so $\hat{R} = \hat{P}$, remainders (of $\widehat{ROQ} - \widehat{OQ}$ and $\widehat{CP} - \hat{C}$).

And $\hat{P} = \hat{D}$,

so $\hat{R} = \hat{D}$,

so $BX > XA$.|

76 Then let AB be bisected at H. Hence it has been demonstrated that if we draw any number of parallel lines to FAM, the circumference of the circle, and they are reflected from circumference FAM so as to produce equal angles, then (the reflected lines) pass between points A and H, and no line
77 among them passes between points B and H.| The nearer one of the lines parallel to AB is to AB, the nearer to point H does its reflection (pass), and the farther one of them is (from AB), the nearer to point A does its reflection (pass).|

78 Let FAΨ be the circumference of a circle in the Prop.
established plane, and let its center be B. Let its radius 3
79 be BA, and let BA be bisected at H.| Let DF cut AB perpendicularly at H. Then each of the arcs DA, AF is a sixth
80 of the circumference.| Let us bisect arc DA at G and arc
81 AF at N, and draw TG parallel to AB, and join GB.| Then when TG is reflected from point G so as to make equal

فزاوية ز مساوية لزاوية د فخطّ بخ اعظم من خطّ خا.

76 فليقسم خطّ اب بنصفين على نقطة ح فقد تبيّن انّه ⟨ان⟩
اخرج الى خطّ ثام المحيط بالدائرة خطوط متوازية كم كانت
وانعطفت من خطّ ثام المحيط بالدائرة واحدثت زوايا متساوية
فانّها تمرّ فيما بين نقطتي اح وليس فيها خطّ [ينعطف فيما]
77 يمرّ فيما بين نقطتي ب ح. وما كان من الخطوط الموازية
لخطّ اب الى خطّ اب اقرب فانّ انعطافه يكون الى نقطة
ح اقرب وما كان منها ابعد فانّ انعطافه يكون الى نقطة ا
اقرب٥

ج

78 ليكن خطّ ثاص محيط بدائرة في السطح الموضوع
وليكن مركزها نقطة ب وليكن ⟨الخطّ الذي يخرج منها
الى الخطّ المحيط ب⟩الدائرة خطّ با وليقسم خطّ با بنصفين
79 على نقطة ح، وليكن خطّ دث قاطعًا خطّ اب ⟨على نقطة
ح⟩ على زوايا قائمة فتكون كلّ واحدة من قوسي دا اث
80 سدس الخطّ المحيط بالدائرة. ولنقسم قوس دا بنصفين على
نقطة ج ولنقسم قوس اث بنصفين على نقطة ن ولنخرج خطّ
81 ثج موازيًا لخطّ اب وليوصل خطّ جب. فخطّ جث اذا

---

76 ثام[1]: بام
77 اقرب[1]: قر ب
79 خطّ اب: لخط اب

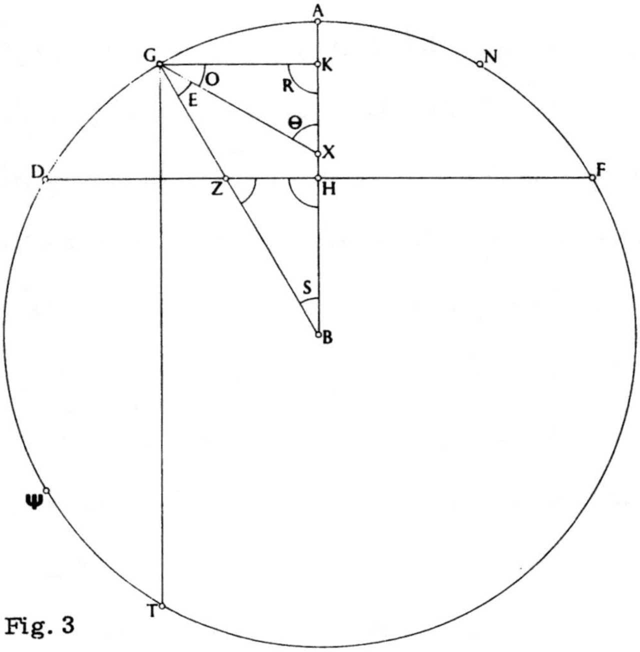

Fig. 3

angles with arc ΨAF, it falls between points A and H, as we
82 showed above.| Let it fall on point X, and let XG make
with GK an angle O, equal to $\hat{S}$. Then it turns out that

$\hat{S}$ is $\frac{1}{3}$ of a right angle,

therefore $\hat{O}$ is $\frac{1}{3}$ of a right angle.|

83 And $\hat{S} = \hat{O}$
and $\hat{R} = \hat{H}$, right angles,
so $\hat{\theta} = \hat{Z}$.

So ΔBHZ is similar to ΔGKX.|
84 And GK = BH, because HA equals both KG and HB,

so GX = BZ.|
85 But GX = BX, because $\hat{S} = \hat{E}$,
so BZ = BX.|

86 But $\hat{S}$ is $\frac{1}{3}$ of a right angle and $\hat{H}$ is a right angle,

so $\hat{Z}$ is $\frac{2}{3}$ of a right angle.

So $BZ^2 = \frac{4}{3} BH^2$,

انعطف من نقطة ج فاحدث مع قوس صات زاويتين
82 متساويتين وقع فيما بين نقطتي اح كما بيّنّا آنفًا. فليقع
على نقطة خ وليحدث خطّ خج مع خطّ جك زاوية مساوية
لزاوية س وهي زاوية ع. فيعرض ان تكون ⟨زاوية⟩ س تلت
83 زاوية قائمة ولذلك تكون زاوية ع تلت زاوية قائمة. وزاوية س
مساوية لزاوية ع وزاوية ز القائمة مساوية لزاوية ح فزاوية ط
84 مساوية لزاوية ز فمثلثا بحز جكخ متشابهان. وخطّ جك
مساو لخطّ بح وذلك انّ خطّ حا مساو لكلّ واحد من
85 خطّي كج حب فخطّ جخ مساو لخطّ بز. ولكنّ خطّ
جخ مساو لخطّ بخ لانّ زاوية س مساوية لزاوية ه فخطّ
86 بز مساو لخطّ بخ. ولكنّ زاوية س تلت زاوية قائمة
وزاوية ح قائمة فزاوية ز تلثا قائمة فمربّع خطّ بز [ايضًا] مرّة
وتلت مثل مربّع خطّ بح فمربّع خطّ بخ ايضًا مرّة
وتلت مثل مربّع خطّ بح. [فمربّع خطّ بخ ايضًا مرّة
87 وتلت مثل مربّع خطّ بح]. فخطّ بح اكثر من ستّة
امثال خطّ حخ وخطّ اح مساو لخطّ بح فخطّ اح اكثر
من ستّة امثال خطّ خح. ولذلك يكون خطّ اخ اكثر من

---

82 لذلك: كدلك؛ تكون: كون

83 وزاوية[1]: فزاويه

86 زاوية[1]: الزاويه؛ وتلت[1]: هملت

so $BX^2 = \frac{4}{3} BH^2$ also.|

87 So BH > 6 HX.
and AH = BH,
so AH > 6 XH,
and therefore AX > 5 XH.|

88 And when the lines parallel to AB which meet arcs GD and NF are reflected at equal angles, they cut AH between
89 points A and X;| as for the section (of AH) beyond X towards H, none of the (above) rays are reflected to it. The rays parallel to AB which meet arc GAN, when reflected at
90 equal angles, cut XH.| The two rays which are reflected from points G and N also cut XH, at X. The nearer one of the other rays (among those reflected from arc GAN) is to A, the nearer its reflection is to H.|

91 So if AB is kept stationary, and arc AD is rotated until it returns to its original position, the resulting figure
92 will be a spherical surface.| If the latter is shaped in brass, and placed facing the sun so that one of the sun's rays passes along AB, then the rays reflected from the surface formed by the rotation of arc GD, when they are re-
93 flected at equal angles, will (all) go to line AX,| while the rays reflected from the whole surface formed by (the rotation
94 of) arc AG will (all) go to line HX,| and the rays coming from the largest circle in that surface, which passes through points G and N, will fall on point X. The rays which are in

88 خمسة امثال خطّ خح. والخطوط الموازية لخطّ اب التي تلقى
قوس جد وقوس نت اذا انعطفت على زوايا متساوية
89 قطعت خطّ اح فيما بين نقطتي اخ. واما خارج نقطة خ
الى ناحية نقطة ح فليس ينعطف ⟨اليه⟩ واحد من الشعاعات.
واما الشعاعات التي تلقى قوس جان وتكون موازية لخطّ اب
اذا انعطفت على زوايا متساوية فانّها تقطع خطّ خح.
90 والشعاعان اللذان ينعطفان من نقطة ج ونقطة ن يقطعان خطّ
خح عند نقطة خ ايضًا. وما كان من الشعاعات الباقية
91 اقرب الى نقطة ا فانّ انعطافه يكون الى نقطة ح اقرب. فاذا
اثبت خطّ ب ا واديرت قوس اد حتى ترجع الى الموضع
الذي منه ابتداءً كان الشكل الذي يحيط به من بسيط الكرة.
92 وان عمل ذلك من صفر ووضع قبالة الشمس حتى يمرّ
شعاع واحد من شعاعات الشمس بخطّ اب فانّ الشعاعات
التي تنعطف من البسيط الذي يحدث من ادارة قوس جد اذا
كانت تنعطف على زوايا متساوية فانّها تصير الى خطّ اخ.
93 واما الشعاعات التي تنعطف من جميع البسيط الحادث عن
94 قوس اج فانّها تصير الى خطّ حخ، والشعاعات التي تخرج

---

90 كان من: كامس

91 ابتداءً: ابتدات

93 قوس: اج add. sed del.

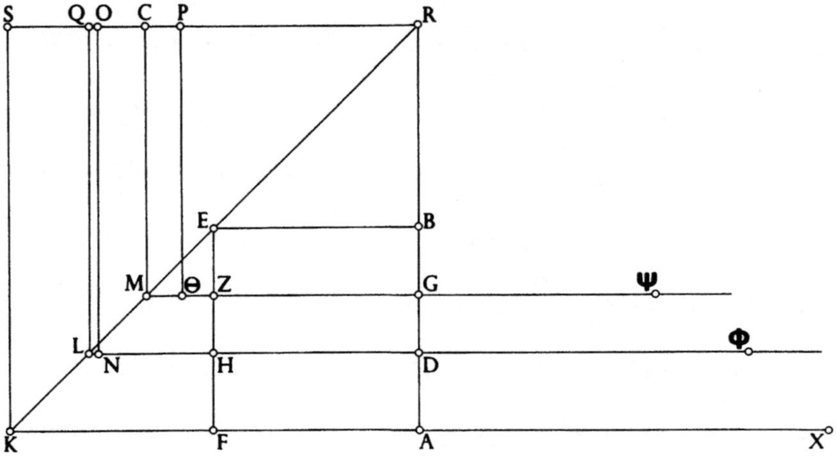

Fig. 4

the vicinity of point A will, after reflection, fall closer
95 to point H.| Thus great heat will result in the space between
points H and X, and it will be nearer to point H than to
point X. Hence there is no advantage in the surface resulting
from arc GD, since no ray from it falls on the area in which
96 the burning takes place.| Therefore whoever wants to make
a burning-mirror from a section of a sphere need use only
the surface produced by arc GA.|

97 How do we shape the curvature of the burning-mirror Prop.
when we want the point at which the burning occurs to be 4
at a given distance from the center of the surface of the
98 mirror?| We draw with a ruler on a given board a line equal
to the distance we want: that is line AB. We make AK twice
BA and erect BE perpendicular and equal to AB; we join
99 EK.| We make AF equal to BE, and join EF: then ABEF is
100 a square, and also EF is equal to FK.| We mark on BA
two points, G, D, and make EZ equal to BG and HE equal

من الدائرة العظيمة التي تمرّ بنقطتي ج ن من ذلك البسيط تقع
عند نقطة خ والشعاعات التي بالقرب من نقطة آ يكون وقوعها
95 اذا انعطفت الى نقطة ح اقرب. فتكون فيما بين نقطتي
ح خ حرارة كبيرة وتكون الى نقطة ح اقرب منها الى نقطة
خ. فليس ينتفع بالبسيط الحادث عن قوس جد [سه] وذلك
انّه لا يقع منه شعاع [سه] على الموضع الذي يكون فيه
96 الاحراق. ولذلك انّما ينبغي لمن اراد ان يعمل مرآة محرقة
من قطعة كرة ان يعمل السطح الذي يحدث من قوس جا
فقط.

د

97 كيف نعمل تقعير المرآة المحرقة اذا اردنا ان تكون
النقطة التي عليها يكون الاحراق على بعد معلوم من وسط بسيط
98 المرآة. وصفنا خطّاً على لوح معلوم على مسطرة مساوياً للبعد
الذي نريد وهو خطّ اب وصيّرنا خطّ اك ضعف خطّ با
وجعلنا ⟨خطّ به⟩ قائمًا على زوايا قائمة ونجعله مساويًا لخطّ
99 اب ونصل هك. ونجعل خطّ ات مساويًا لخطّ به ونصل
خطّ هت فيصير ابهت مربّعًا وايضًا فانّ هت يكون مساويًا
100 لخطّ تك. ونتعلّم على خطّ با نقطتي ج د ونجعل خطّ
هز مساويًا لخطّ بج وخطّ حه مساويًا لخطّ بد ⟨ونصل

---

94 التي[3]: ىمر add. sed del.; انعطفت: انعطف
97 المحرقة: المعرفه
100 حه: ىه; نخرجهما: نخرجها; بخطّ: على

to BD. We join ZG, HD, and produce them on both sides:
101 let them meet EK in M, L.| Then if, with A as center and
GM as radius, we draw a circle, it cuts GM: let it cut it
102 in θ.| Then we continue to draw it in the same way until it
cuts it in Ψ. Again, if, with center A and radius DL, we
103 draw a circle, it cuts DL: let it cut in N.| Then we continue
to draw it about center A until it cuts it (DL) again in Φ.
Then we draw AX as an extension in a straight line of KA
and make it (AX) equal to it (KA). Then points K, N, θ, B,
Ψ, Φ, X lie on a parabola.|

104 For we produce AB to R, letting BR equal AB; let us
draw RS perpendicular to AB and equal to KA, and join SK,
and draw from points L, M, θ, N to line RS perpendiculars
105 LQ, MC, NO, θP.| Then when KE is produced in a straight
line it passes through R.

So QL = LD
and MC = MG, because KER is a diagonal of square AS.|
106 But LD = NA
and MG = θA,

خطّي زج حد> ونخرجهما في كلتي الجهتين وليلتقيا بخطّ
101 هك على نقطتي م ل. واذا جعلنا نقطة ا مركزًا وادرنا ببعد
102 جم دائرة قطعت خطّ جم فلتقطعه على نقطة ط. ثمّ نديرها
على هذا المثال حتى تقطعه على نقطة ص. وايضًا فانّا جعلنا
نقطة ا مركزًا وادرنا دائرة ببعد مثل دل قطعت خطّ دل
103 فلتقطعه على نقطة ن. ثمّ نديرها والمركز على نقطة ا حتى
تقطعه ايضًا على نقطة ض. ثمّ نخرج خطّ اخ على استقامة
خطّ كا ونجعله مساويًا له. فتكون نقط ل ن<ط>
104 ب ص ض خ على قطع المخروط القائم الزاوية. وذلك انّا
نخرج خطّ با الى نقطة ر وليكن خطّ بر[ه] مساويا لخطّ
اب ولنخرج خطّ رس وليكن مساويًا لخطّ كا وليكن على
خطّ اب على زوايا قائمة وليوصل خطّ سك ولنخرج من نقط
105 ل م ط ن الى خطّ رس اعمدة لق مش نع طف. فخطّ
كه اذا اخرج على استقامة وقع على نقطة ر. فيكون خطّ قل
مساويًا لخطّ لد وخطّ مش مساو لخطّ مج وذلك انّ خطّ
106 كهر هو قطر مربّع اس. وحطّ لد مساو لخطّ نا وخطّ

---

101 فلتقطعه: فلقطه

102 فلتقطعه: فلقطه

103 تقطعه: نقطه

106 وخطّ[1]: فخط

and LQ = LD also
and MC = MG.|
107 So AN = LQ = NO
and Aθ = MC = θP.
And AK = KS
and AR is bisected at B.|

108 Since that is so, points B, θ, N, K lie on a parabola, as we shall prove subsequently. Similarly points Ψ, Φ, X also (lie on the parabola).|

109 So if we mark numerous points on AB, and draw through them lines parallel to AK, and mark on the lines points corresponding to the other points (i.e. θ, N etc.), and bend
110 along the resultant points a ruler (?)made of horn,| fastening it so that it cannot move, then draw a line along it and cut the board along that line, then shape the curvature of the
111 figure we wish to make to fit that template,| the burning from that surface will occur at point A, as was proved in the first proposition.|

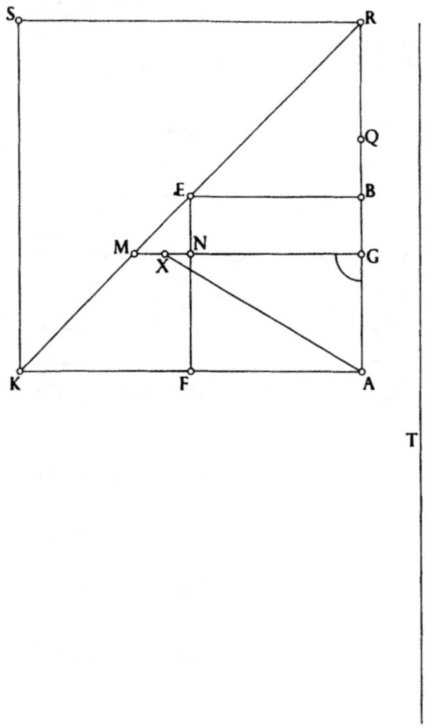

Fig. 5

مج مساو لخطّ طا وخطّ لق ايضًا مساو لخطّ لد وخطّ
107 مش مساو لخطّ مج، فخطّ ان مساو لخطّ لق الذي هو
مثل خطّ نع وخطّ اط مساو لخطّ مش الذي هو مثل خطّ
〈ط〉ف. وخطّ اك مساو لخطّ كس وقد قسم خطّ ار[ه]
108 بنصفين على نقطة ب. واذا كان ذلك كذلك فانّ نقط
ب ط ن ك على قطع المخروط القائم الزاوية كما نبيّن فيما
109 بعد وكذلك ايضًا نقط ص ض خ. فاذا تعلّمنا على خطّ اب
نقطًا كثيرة واخرجنا منها خطوطًا موازية لخطّ اك وتعلّمنا في
〈الخطوط نقطًا يكون حالها حال〉 النقط الآخر وعطفنا على
110 النقط التي تحدث مسطرة من قرون، واثبتناها حتى لا يمكن
ان تتحرّك وخطّطنا عليها خطًّا وقطعنا على ذلك الخطّ اللوح
وجعلنا تقعير الشكل الذي نريد [ح] عمله على ذلك القالب،
111 فانّ الاحراق يكون من السطح على نقطة ا وتبيّن ذلك في
الشكل الاوّل.

112 وليكن قطر مربّع اس خطّ رك وليقسم خطّ ار ٥
بنصفين على نقطة ب ولتتعلّم نقطة ما فيما بين نقطتي ا ب
113 وهي نقطة ج. وليجاز على نقطتي ب ج خطّان موازيان لخطّ
اك وهما ب ه [ب ه] ج م وعلى نقطة ه خطّ مواز لخطّ اب

---

109 اك: ان

110 ان: لا: ذلك[2]: حك

112 Let the diagonal of square AS be line RK, and let us Prop.
bisect AR at B, and mark some point, G, between A and 5
113 B.| Let two lines parallel to AK, BE and GM, pass through points B and G, and let a line parallel to AB, EF, pass through point E. Then

MG - MN = BE.|

114 And because of what was stated in the proposition preceding this one, AE is a square,

so BE = AB;
and EN equals both NM and BG,
therefore MN = BG.
So MG - BG = AB.|
115 Let BQ be equal to GB.
Then QA = MG.

So the circle constructed on center A with radius equal
116 to MG passes through Q.| Then I say that it (the circle) cuts line MG between points M and G. For if it were to pass through M or fall beyond M, its radius would be longer than GM, since $\hat{G}$ is right, and that is impossible, since we
117 have made it (the radius) equal to it (GM).| So the circumference of the above-mentioned circle cuts MG between points M and G. Then let it cut it in X, and let us draw a line, XA, joining points X and A. Then

AX = MG.|

118 And we have shown that AQ also equals MG,

وهو خطّ هت. فخطّ مج يفضل على خطّ به بخطّ من.
114 فمن اجل ما قدّمنا في الشكل الذي قبل هذا يصير اه مربّعًا
وخطّ هب مساو لخطّ با. وخطّ هن مساو لكلّ واحد من
خطّي نم بج فخطّ من مساو لخطّ بج. فخطّ مج
115 يفضل على خطّ اب بخطّ بج. ليكن خطّ بق مساويًا
لخطّ جب. ولذلك يصير خطّ قا مساويًا لخطّ مج. فالدائرة
التي تعمل على مركز ا وببعد مساو لخطّ مج تمرّ بنقطة
116 ق. فاقول انّها تقطع خطّ مج فيما بين نقطتي م ج وذلك
انّها ان مرّت بنقطة م او وقعت فوق نقطة م صار الخطّ الذي
يخرج من مركزها الى الخطّ المحيط بها اطول من خطّ
جم لانّ زاوية ج قائمة وذلك غير ممكن وذلك انّا كنّا
117 جعلنا⟨ه⟩ مساويًا له. فالخطّ المحيط بالدائرة التي ذكرنا يقطع
خطّ مج فيما بين نقطتي م ج. فليقطعه على نقطة خ
وليوصل بين نقطة خ وبين نقطة ا بخطّ خا فخطّ اخ مساو
118 لخطّ مج. وقد بيّنّا انّ خطّ اق ايضًا مساو لخطّ مج فخطّ
خا مساو لخطّ اق. ولانّ خطّ قب مساو لخطّ بج صار
ما يجتمع من ضرب خطّ اب في خطّ بج اربع مرّات مع
المربّع الكائن من خطّ جا مساويًا للمربّع الكائن من خطّ

---

113 اب: اك

115 ليكن: لكن

118 مساويًا: مساو

so XA = AQ.

And since QB = BG,

$4\,(AB.BG) + GA^2 = AQ^2$.|

119 But $AQ^2 = AX^2$,

and $XG^2 + GA^2 = AX^2$; because $\hat{G}$ is right.|

120 So $4\,(AB.BG) + GA^2 = XG^2 + GA^2$.|

121 And when we eliminate GA^2, common,

$4\,(AB.BG) = XG^2$, remainders.|

122 And $4\,(AB.BG) = BG.4AB$.|

123 Let $4AB = T$.

Then $XG^2 = BG.T$.

And also $AK^2 = AB.T$, because SA is a square, and AR is
124 bisected at B.| So the parabola which passes through points B and K will also pass through X, and the parameter of the squares on the ordinates is line T, which is what we wanted to prove.|

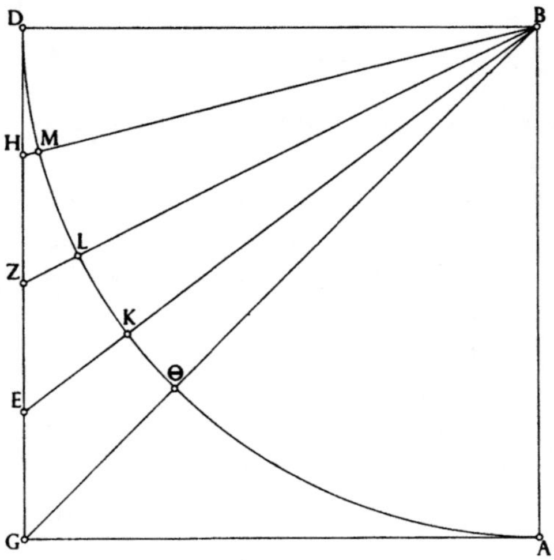

Fig. 6a

119 اق. والمربّع الكائن من خطّ اق مساو للمربّع الكائن من
خطّ اخ والمربّعان الكائنان من خطّي خج جا ⟨مساويان⟩
120 للمربّع الكائن من خطّ اخ لانّ زاوية ج قائمة. فالمجتمع من
ضرب خطّ اب في خطّ بج اربع مرّات مع المربّع
الكائن من خطّ جا مساو للمربّعين الكائنين ⟨من خطّي⟩
121 خج جا. واذا اسقط المربّع الكائن من خطّ اج المشترك
صار الباقي وهو المجتمع من ضرب خطّ اب في خطّ جب
اربع مرّات مساويًا للباقي وهو المربّع الكائن من خطّ
122 خج. والمجتمع من ضرب خطّ اب في خطّ [ا] بج اربع
مرّات مساو للمجتمع من ضرب خطّ بج في اربعة امثال
123 خطّ اب. فليكن اربعة امثال خطّ اب خطّ ت. فالمربّع
الكائن من خطّ خج مساو للمجتمع من ضرب خطّ بج
في خطّ ت. والمربّع الكائن من خطّ اك ايضًا مساو
للمجتمع من ضرب خطّ اب في خطّ ت وذلك انّ سا
124 مربّع وقد قسم خطّ ار بنصفين على نقطة ب. فالقطع القائم
الزاوية الذي يمرّ بنقطتي ب ك يمرّ بنقطة خ ايضًا ويصير
[ويصير] الخطّ الذي تقوى عليه الاعمدة خطّ ت وذلك ما اردنا
ان نبيّن٥

---

121 مساويًا: مساو

123 فليكن: ولكن

125 Then after that Diocles proves that equal quantities situated on a straight line are subtended by unequal (angles), and that the greatest (of the latter) is the one nearest the perpendicular drawn from the eye to that line. We must prove Prop. 6
126 that.| So let a square ABGD be drawn, and let BG be joined. On center B, with radius BD, draw $\widehat{D\theta A}$ cutting BG in θ. It is clear that it will pass through A, and that $\widehat{D\theta}$ will be
127 equal to $\widehat{\theta A}$.| Then let GD be divided into any number of equal lines GE, EZ, ZH, HD. Join EB, ZB, HB, and let them cut
128 $\widehat{D\theta}$ in K, L, M.| Then it is clear that angles DBM, MBL, LBK, KBθ will be unequal, so the quantities DH, HZ, ZE, EG are subtended by unequal (angles). And line GD is longer than arc Dθ, because GD plus GA is longer than the whole arc
129 DθA.| So let there be a template equal (in size and shape) to the figure bounded by arc θLD and lines θB and BD, and let a sheet be bent along $\widehat{\theta LD}$, and let points K, L, M be
130 marked on it.| Then let the sheet be flattened out and let it

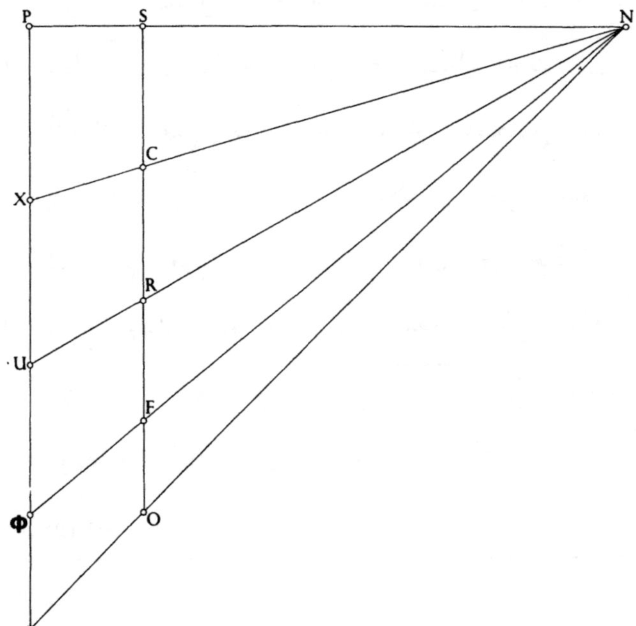

Fig. 6b

و

125 تمرّ من بعد ذلك يبيّن [ديو] ذيوقليس انّ الاقدار
المتساوية الموضوعة على خطّ مستقيم ترى غير متساوية واعظمها
الذي يقرب من العمود الذي يخرج من البصر الى ذلك الخطّ
126 وينبغي ان نبيّن ذلك. فليوضع مربّع اب جد وليوصل خطّ
بج ولترسم على مركز ب وببعد بد قوس دطا
⟨ولتقطع خطّ بج على نقطة ط⟩. وهو بيّن انّها تمرّ بنقطة
127 ا وتصير قوس دط مساوية لقوس طا. فليقسم خطّ جد لخطوط
متساوية كم كانت وهي خطوط جه هز زح حد وليوصل
خطوط هب زب ح⟨ب⟩ ولتقطع قوس دط على نقط
128 ك ل م⟩. فهو بيّن انّ زوايا دبم ⟨مبل لبك
كبط غير متساوية⟩ فاقدار دح حز زه هج ترى غير
متساوية. وخطّ جد اطول من قوس دط لانّ خطّي جد جا
129 اطول من جميع قوس دطا. فليكن قعر ما مساو للشكل
الذي تحيط به قوس طلد وخطّي طب بد ولتعطف على
130 قوس طلد صفيحة ما ولتتعلّم عليها نقط ك ل م. ثمّ لتبسط
الصفيحة ولتكن سع ولتكن النقط التي كانت تعلّمت عليها

---

126 مركز ب: جد add. sed del.
128 غير[2]: عفر: جا: ما
129 ولتتعلّم: ولعمل
130 لتبسط: بسط

(i.e. the part of it corresponding to arc θLD) be SO, and let the points marked on it be C, R, F. Let us draw SN from S perpendicular to SO, and make SN equal to SO. Let us join NO, and produce it in the other side to Q, and make
131 NP equal to GD.| Then it (NP) is longer than NS, because we said in the preceding section that

$$GD > \widehat{\theta LD},$$
$$\text{and } \widehat{\theta LD} = NS.$$

Let there be drawn through point P a line PQ parallel to SO, and let NC, NR, NF be joined and produced to X, U and Φ (respectively).|

132 Then

$$\widehat{GBD} : \widehat{HBD} = \widehat{\theta LD} : \widehat{MD} = SO : SC = PQ : PX = GD : PX.|$$

133 But GD is subtended by $\widehat{GBD}$, so PX is subtended by $\widehat{HBD}$.
134 And DH is also subtended by HBD.| Similarly ZH also is subtended by (an angle) equal to (that subtending) XU, EZ by (one) equal to (that subtending) UΦ, and GE by (one) equal to (that subtending) ΦQ. So lines DH, HZ, ZE, EG are subtended by (angles) equal to those subtending PX, XU, UΦ, ΦQ.|
135 And it is clear that each of the lines PX, XU, ΦU, ΦQ is

نقط ش ر ت ولنخرج من نقطة س خطّ س ن على خطّ س ع
على زوايا قائمة ونجعل خطّ س ن مساويًا لخطّ س ع ونصل
خطّ ن ع ونخرجه الى الجهة الاخرى الى نقطة ق ونجعل خطّ
131 ن ف مساويًا لخطّ ج د. فهو يكون اطول من ن س وذلك انّا
قد قلنا فيما تقدّم انّه اطول من قوس ط ل د التي هي مثل
خطّ ن س. وليجاز على نقطة ف خطّ مواز لخطّ س ع وهو خطّ
ق ف ولتوصل خطوط ن ش ن ر ن ت ولتخرج الى نقط
132 خ ذ ض. فنسبة زاوية ج ب د الى زاوية ح ب د كنسبة قوس
ط ل د الى قوس م د التي هي مثل نسبة خطّ س ع الى خطّ
س ش التي هي مثل نسبة خطّ ق ف الى خطّ ف خ التي هي
133 مثل نسبة خطّ ج د الى خطّ ف خ. ويرى خطّ ج د من زاوية
ج ب د فخطّ ف خ يرى من زاوية ح ب د ⟨وخطّ د ح ايضًا
134 يرى من زاوية ح ب د⟩. وكذلك ايضًا يرى خطّ ز ح ايضًا
مثل خطّ خ ذ وخطّ ه ز يرى مثل خطّ ذ ض وخطّ ج ه يرى
مثل خطّ ض ق. فخطوط د ح ح ز ⟨ز⟩ه ⟨ه⟩ج ترى مثل ما
135 ترى خطوط ف خ خ ذ ذ ض ⟨ض⟩ق، وهو بيّن انّ خطوط
ف خ خ ذ ض ذ ⟨ض⟩ق بعضها اطول من بعض لانّ قسي
د م م ل ل ك ك ط ايضًا بعضها اطول من بعض لانّ
الزوايا التي عند المركز ايضًا بعضها اعظم من بعض٥

---

131 ن ر: ن ت

greater than the next, since each of the arcs DM, ML, LK, KΘ too is greater than the next, because each of the angles at the center too is greater than the next.|

136 Archimedes proved in his book *On the Sphere and* Prop.
Cylinder that every segment of a sphere is equal to the cone 7
137 whose base is the same as that of the segment| and whose height is (equal to) a line whose ratio to the perpendicular between the vertex and the base of the segment equals the ratio of the sum of the radius of the sphere and the perpendicular, i.e. height, of the other segment of the sphere
138 to the perpendicular of that second segment.| For example, let there be a sphere, ABG, cut by a plane, namely the plane of the circle with diameter GD, and let AB be the
139 diameter of circle ABG, and point E its center.| We set

$$HZ : ZB = (EA + AZ) : ZA,$$

and, by a similar construction, we derive the ratio ΘZ : ZA.|
140 Then it has been proven that segment GBD of the sphere is equal to the cone whose base is the circle of diameter GD and whose axis is ZH, and that segment GAD is equal to the cone whose base is the same circle and whose axis is ΘZ.|
141 Having proved that, he (Archimedes) wanted to cut the given sphere by a plane so that the two segments of the
142 sphere bear a given ratio to one another.| So he said that the ratio of the cone whose base is the circle with diameter GD and whose height is ZΘ to the cone whose base is the

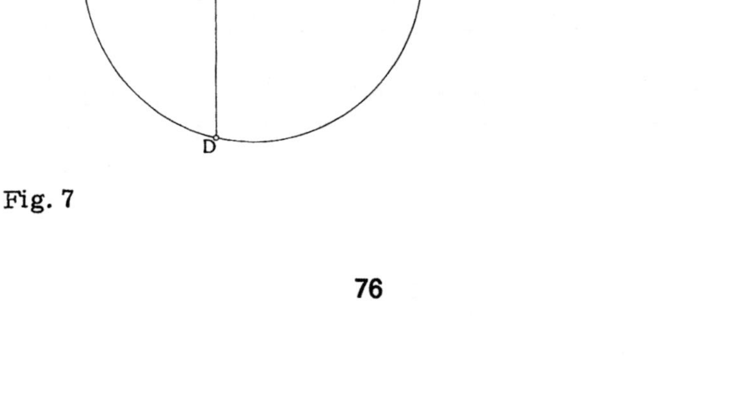

Fig. 7

136 وقد بيّن ارشميذس في القول في الكرة والاسطوانة انّ ز
كلّ قطعة كرة فهي مساوية للمخروط الذي قاعدته قاعدة القطعة،
137 وارتفاعه خطّ ما نسبته الى العمود الذي [الا] يخرج من رأس
القطعة الى قاعدتها مثل نسبة نصف قطر الكرة وعمود القطعة
الاخرى من الكرة الذي هو ارتفاعها جميعًا الى عمود تلك القطعة
138 الثانية. مثال ذلك ⟨لتكن⟩ كرة ابج ولتقطع بسطح ما
وهو سطح الدائرة التي يكون قطرها جد وليكن اب قطر دائرة
139 ابج ونقطة ه مركزها. ونجعل نسبة خطّ حز الى خطّ زب
كنسبة خطّي ها از مجموعين الى خطّ زا وبمثل هذا العمل
140 ايضًا نستخرج نسبة خطّ طز الى خطّ زا. فقد تبيّن انّ قطعة
جبد من الكرة مساوية للمخروط الذي قاعدته الدائرة التي يكون
قطرها جد وسهمه ⟨خطّ زح وقطعة جاد مساوية للمخروط الذي
141 قاعدته هذه الدائرة بعينها وسهمه⟩ خطّ طز. فلمّا تبيّن له ذلك
اراد ان تقطع الكرة المعلومة بسطح ما حتى تكون لقطعتي
142 الكرة احديهما الى الاخرى نسبة مثل نسبة معلومة. فقال انّ
نسبة المخروط الذي قاعدته الدائرة التي على قطر جد وارتفاعه
خطّ طز الى المخروط الذي قاعدته هذه الدائرة بعينها وارتفاعه

---

136 قاعدته: فاعده

137 الى: التي

141 المعلومة: المعموله

143 same circle and whose height is ZH is given,| and it is equal to the ratio of θZ to ZH, since it has been proven that cones on equal bases have to one another the ratio of their heights.|
144 So the ratio of θZ to ZH is given, and since

$$\Theta Z : ZA = (EB + BZ) : BZ,$$

dirimendo,

$$\Theta A : AZ = BE : BZ.|$$

145 And similarly we prove also that

$$HB : BZ = EA : AZ.$$
But EA = BE.|

146 So this problem has turned out to be as I describe, namely:
If line AB and two points, A, B, be given in position,
147 and line EB be given in magnitude,| how to divide AB at point Z and add to it θA and BH such that the ratio of θZ to ZH be given, and furthermore the ratio of θA to ZA be
148 equal to the ratio of the given line, EB, to ZB,| and furthermore the ratio of HB to ZB be equal to the ratio of the same given line, EB, to ZA. We shall expound that (problem)
149 in what follows;| for Archimedes, having explained the abovementioned (problem) in a manner longer than ours, arrived at another problem which he did not solve in his book *On the Sphere and Cylinder*.|

143 خطّ زح معلومة، ⟨وهي⟩ كنسبة خطّ طز الى خطّ زح وذلك
لانّه قد تبيّن انّ المخروطات التي تكون على قواعد متساوية
فانّ نسبة بعضها الى بعض كنسبة ارتفاع بعضها الى بعض.
144 فنسبة خطّ طز الى خطّ زح معلومة ولانّ نسبة خطّ طز الى
خطّ زا كنسبة خطّي ەب بز جميعًا الى خطّ بز تكون اذا
فصّلنا نسبة خطّ طا الى خطّ از كنسبة خطّ بە الى خطّ
145 بز، وكذلك ايضًا نبيّن انّ نسبة خطّ حب الى خطّ بز
[وكذلك ايضًا نبيّن انّ نسبة خطّ حب الى خطّ بز]
كنسبة خطّ ەا الى خطّ از. وخطّ ەا هو مثل خطّ بە.
146 فصارت هذه المسئلة على ما اصف وهو انّه اذا كان خطّ
اب معلوم الوضع وكانت نقطتا اب معلومتي الوضع
147 وكان خطّ ەب معلوم القدر، كيف نقسم خطّ اب على نقطة
ز ونضيف اليه خطّي طا بح حتى تكون نسبة خطّ طز الى
خطّ زح معلومة وتكون مع ذلك نسبة خطّ طا الى خطّ زا
148 كنسبة الخطّ المعلوم الذي هو ەب الى خطّ زب، وتكون مع
ذلك نسبة خطّ حب الى خطّ زب كنسبة ذلك الخطّ المعلوم
149 نفسه الذي هو ەب الى خطّ زا وسنبيّن ذلك فيما بعد. وذلك
انّ ارشيميذس لمّا بيّن ما قلناه بوجه اطول من هذا صار
الى مسئلة اخرى لم يبيّنها في كتابه في الكرة

---

147 ونضيف: وىصف

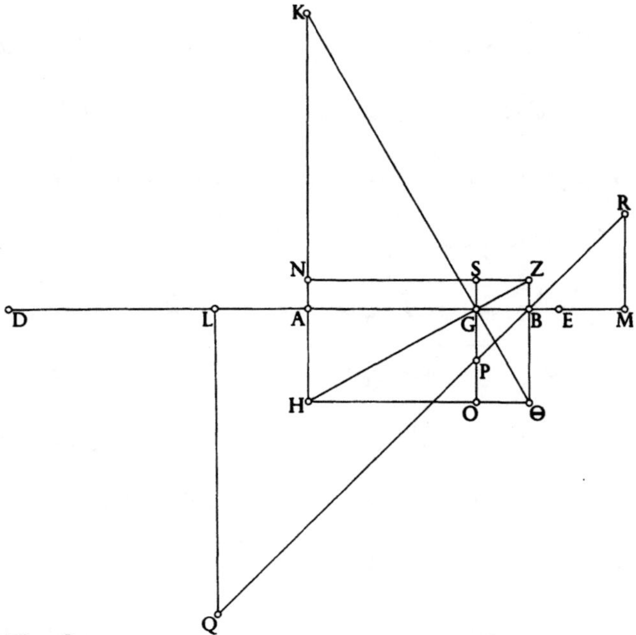

Fig. 8

150 When line AB is given in position, and two points, A, B, are given in position, how may we cut AB at G in such a way that when lines DA, BE, are added to it the result is Prop. 8
151 that the ratio of GD to GE is equal to a given ratio,| and furthermore that the ratio of DA to AG is equal to the ratio of some given line to GB, and that the ratio of EB to BG is equal to the ratio of that given line to GA.|
152 Let angles GAH, GBθ be right, and let us make each of lines AH, Bθ equal to the given line. Let HG, Gθ be joined
153 and produced to points Z, K.| Then since

DA : AG = the given line θB : BG = KA : AG,
KA = AD.

Similarly we prove also that

ZB = BE.|

154 And it is clear that the rectangle Aθ is given, for both AB
155 and AH are given.| And since DG : GE is given,

〈والاسطوانة〉.

ح

150 اذا كان خطّ اب معلوم الوضع وكانت نقطتا ا ب معلومتي الوضع كيف نقسم خطّ اب على نقطة ج قسمة اذا اضفنا اليه خطّي دا به صارت نسبة خطّ جد الى خطّ جه
151 كنسبة معلومة، وصارت ايضًا نسبة خطّ دا الى خطّ اج كنسبة خطّ ما معلوم الى خطّ جب وصارت نسبة خطّ هب الى خطّ
152 بج كنسبة ذلك الخطّ المعلوم الى خطّ جا. فلتكن زاويتا جاح جبط قائمتين ونجعل كلّ واحد من خطّي اح بط مساويًا للخطّ المعلوم وليوصل خطّا حج جط وليخرجا الى
153 نقطتي زك. فلانّ نسبة خطّ دا الى خطّ اج كنسبة الخطّ المعلوم الذي هو خطّ طب الى خطّ بج وهذه النسبة مثل نسبة خطّ كا الى خطّ اج يكون خطّ كا مساويًا لخطّ اد
154 وكذلك ايضًا نبيّن انّ خطّ زب مساو لخطّ به. وهو بيّن انّ سطح اط المتوازي الاضلاع معلوم وذلك انّ كلّ واحد
155 من خطّي اب اح معلوم. ولانّ نسبة خطّ دج الى خطّ جه معلومة تكون نسبة خطّي كا اج جميعًا الى خطّي بز بج

---

150 الوضع: بالوصع؛ خطّ جد: خطي جد (ي .del)
152 جبط: جبر
153 اج: اب؛ طب: ططب (ط[2] .del)؛ كا[1]: دا؛ كا[2]: طا؛ مساو: مساويا

$(KA + AG) : (BZ + BG)$ is given,|

156 so $(KA + AG) . (ZB + GB) : (ZB + BG)^2$ is given.|

157 And since

$$(AH + AG) : (GB + BZ) = (KA + AG) : (\theta B + BG),$$

for each of these ratios is equal to AG : BG,|

158 $(HA + AG) . (\theta B + BG) = (KA + AG) . (ZB + BG)$.|

159 But $(KA + AG) . (ZB + BG) : (ZB + BG)^2$ is given.|

160 So we make AL equal to HA, and BM equal to θB. Then
161 points L and M are given,| and it is clear that if L falls between D and A, M will fall beyond E, since we have proven that

$$DG . GE = LG . GM.|$$

162 The result is that $LG . MG : (ZB + BG)^2$ is given.|
163 Let there be drawn through Z a line, ZN, parallel to

156 جميعًا معلومة، فنسبة المجتمع من ضرب خطّي كا اج مجموعين
في خطّي زب ⟨ب⟩ج مجموعين الى المربّع الكائن من خطّي
157 زب بج مجموعين معلومة. ولانّ نسبة خطّي اح اج مجموعين
الى خطّي جب بز مجموعين كنسبة خطّي كا اج مجموعين
الى خطّي طب بج مجموعين وذلك انّ كلّ واحدة من
هاتين النسبتين هي ⟨مثل⟩ نسبة خطّ اج الى خطّ بج،
158 يكون المجتمع من ضرب خطّي حا اج مجموعين في خطّي
طب بج مجموعين مساويًا للمجتمع من ضرب خطّي كا
159 اج مجموعين في خطّي زب بج مجموعين. ولكنّ نسبة
المجتمع من ضرب خطّي كا اج مجموعين في خطّي زب
بج مجموعين الى المربّع الكائن من خطّي زب بج
160 مجموعين معلومة. فنجعل خطّ ال مساويًا لخطّ حا ونجعل خطّ
161 بم مساويًا لخطّ طب. فنقطتا ل م معلومتان، فهو بيّن انّ
نقطة ل اذا كانت فيما بين نقطتي دا تصير نقطة م خارج
نقطة ه لانّا قد بيّنّا انّ المجتمع من ضرب دج في جه مساو
162 للمجتمع من ضرب لج في جم. فتصير نسبة المجتمع من
ضرب لج في خطّ مج الى المربّع الكائن من خطّي زب
163 بج مجموعين معلومة. وليجاز على نقطة ز خطّ مواز لخطّ اب

161 لج: ببج

162 نسبة: لنا

AB, and let us draw through G a line, SGO, cutting AB at
164 right angles;| let us make GP equal to BG, and let BP be
joined and produced to points Q, R. Then line QR is given
in position, since BG : GP is given. So let two perpen-
diculars be drawn to DM from points L and M, namely LQ
165 and MR.| Then

$$LG : GM = QP : PR = QP^2 : (QP.PR).|$$

166 So, *permutando*,

$$LG^2 : QP^2 = (LG.GM) : (QP.PR).|$$

167 And $LG^2 : QP^2$ is given, for $GB^2 : BP^2$ is given.|

168 So (LG.GM) : (QP.PR) is given,
and (LG.GM) : $(ZB + BG)^2$ is given,
and $(ZB + BG)^2 = SP^2$.|

169 So (QP.PR) : SP^2 is given, and $\widehat{SPR}$ is given, since it is
half a right angle, and points Q and R are given.|

وهو خطّ زن ونجيز على نقطة ج خطًّا يقطع خطّ اب على
164 زوايا قائمة وهو سجع، ونجعل خطّ جف مساويًا لخطّ بج
وليوصل خطّ بف وليخرج الى نقطتي ق ر. فخطّ قر
معلوم الوضع لانّ نسبة خطّ بج الى خطّ جف معلومة.
فليخرج عمودان على دم من نقطتي ل م وهما لق مر.
165 فنسبة خطّ [لج في جم ونسبة] لج الى جم كنسبة خطّ
قف الى خطّ فر التي هي مثل نسبة المربّع الكائن من
خطّ قف الى المجتمع من ضرب خطّ قف في خطّ فر.
166 واذا بدّلنا تكون نسبة المربّع الكائن من خطّ لج الى
المربّع الكائن من خطّ قف كنسبة المجتمع من ضرب
167 لج في جم الى المجتمع من ضرب قف في فر. ونسبة
المربّع الكائن من لج الى المربّع الكائن من قف معلومة
لانّ نسبة المربّع الكائن من خطّ جب الى المربّع الكائن
168 من خطّ بف معلومة. فنسبة المجتمع من ضرب لج في
جم الى المجتمع من ضرب قف في فر معلومة ونسبة
المجتمع من ضرب لج في جم الى المربّع الكائن من
خطّي زب بج مجموعين الذي هو مثل المربّع الكائن من
169 خطّ سف معلومة. ⟨فنسبة المجتمع من ضرب خطّ قف في

---

163 زن: ون؛ ونجيز: ويحفر

164 معلومة: معلوم

170 So S lies on the perimeter of an ellipse given in po-
sition.
And rectangle Aθ is equal to rectangle SH, since ZH
is a diagonal of rectangle Nθ. So

$$\theta H . \theta B = SN . SO.$$

And lines KH, Hθ are given in position, and point B is given.|
171 So if we construct a hyperbola with asymptotes KH, Hθ,
and passing through the given point B, it will also pass
through point S.|
172 So point S is on the perimeter of a hyperbola given
in position, and is also on the perimeter of an ellipse
173 given in position.| And SG is perpendicular to AB, so
point G is given. And the ratio of EB to BG equals the
ratio of the given line, which equals AH, to the given
174 line AG.| So the ratio of BE to BG is given, so point E
also is given. And similarly point D also is given, and the
synthesis of that (problem) is obvious.|
175 We want to show how to find a line equal to one and a Prop.
fraction times a given line. Let the given line be DE: we 9
want to find a line which is $1\,^1/8$ times the given line, or
(?) $^8/7$ths of it.|

خطّ فر الى المربّع الكائن من خطّ سف معلومة.⟩ وزاوية
سفر معلومة وذلك انّها نصف زاوية قائمة ونقطتا ق ر[ه]
170 معلومتان. فنقطة س هي على خطّ محيط بقطع ناقص معلوم
الوضع وسطح اط مساو لسطح سح لانّ زح قطر سطح
نط. فالمجتمع من ضرب طح في طب مساو للمجتمع من
ضرب سن في سع. وخطّا كح حط معلوما الوضع ونقطة
171 ب معلومة. فان عملنا قطعًا زائدًا يكون خطّا كح حط الخطّين
اللذين لا يلقيانه ويمرّ بنقطة ب المعلومة فانّه يمرّ بنقطة س.
172 فنقطة س هي على خطّ محيط بقطع زائد معلوم الوضع وهي
173 ايضًا على ⟨خطّ⟩ محيط بقطع ناقص معلوم الوضع. وخطّ
سج عمود على خطّ اب فنقطة ج معلومة. ونسبة خطّ هب الى
خطّ بج كنسبة الخطّ المعلوم الذي هو مثل خطّ اح الى
174 خطّ اج المعلومة. فنسبة خطّ هب الى خطّ بج معلومة فنقطة ه
ايضًا معلومة. وكذلك تعلم نقطة [نقطة] د وتركيب ذلك بيّن ه
175 نريد ان نبيّن كيف نجد خطًّا يكون مثلًا وجزءًا لخطّ ط
معلوم. فليكن الخطّ المعلوم خطّ ده ونريد ان نجد خطًّا ما

---

170 نط: رط

171 المعلومة: معلومه

173 وخطّ: فخطه: المعلومة: معلومه

174 ذلك: دكن

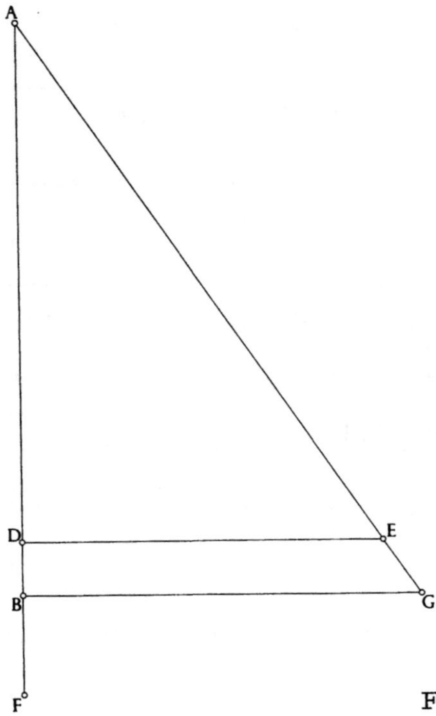

Fig. 9

176 Let DA be perpendicular to DE, and let us produce AD
177 in a straight line to F, and mark any point B on DF.| Let
DA be 8 times DB. Let us produce AE in a straight line
178 to G, and draw GB parallel to DE.| Then since

AD = 8 DB
BA = 1 $^{1}/8$ AD.
So BG = 1 $^{1}/8$ DE also.

So we have found the line we wanted, namely line BG.|
179 So if we want to find another such line such that, when
180 BG is given, BG is 1 $^{1}/8$ times that line,| we draw BA perpendicular (to BG), as in the diagram, and mark point D
181 anywhere on it, and make AD 8 times DB,| and draw AG
and make DE parallel to BG. Then BG is 1 $^{1}/8$ times DE.|
182 So if we want it to be 1 $^{1}/7$ times (DE), we make AD
7 times DB, and carry out all the other operations exactly
as we did (before).|
183 Let the given line be BG: we want to find a line, e. g.
line DE, such that BG is one and a composite fraction times

يكون مرّة وثمن مثل الخطّ المعلوم او ثماني †جزء وضعه
176 المواضع†. فليكن خطّ دا قائمًا على خطّ ده على زوايا قائمة
وليخرج خطّ اد على استقامة ⟨الى نقطة ث⟩ ولنتعلّم على خطّ
177 دث نقطة كيف ما وقعت وهي نقطة ب، وليكن خطّ دا
ثمنية امثال خطّ دب وليخرج اه ⟨على استقامة الى نقطة⟩ ج
178 ولنخرج خطّ جب موازيًا لخطّ ده. فلانّ خطّ اد ثمنية
امثال خطّ دب يكون خطّ با مرّة وثمنًا مثل خطّ اد. فخطّ
بج ايضًا مثل خطّ ده وثمنه. فقد وجدنا الخطّ الذي اردنا
179 وهو خطّ بج. فان اردنا ان نجد خطًّا آخر اذا كان خطّ
ب[فو] ج معلومًا حتى يكون خطّ بج مرّة وثمن مثل ذلك
180 الخطّ، فانّا نخرج خطّ با على زوايا قائمة على ما وصفنا
ونتعلّم عليه نقطة د كيف ما وقعت ونجعل اد ثمنية امثال
181 دب، ونخرج خطّ اج ونجعل ده موازيًا لخطّ بج فيصير
182 خطّ بج مرّة وثمنًا خطّ ده. فان اردنا ان يكون مرّة وسبعًا
صيّرنا خطّ اد سبعة اضعاف خطّ دب ونعمل تلك الاشياء
183 الباقية التي عملنا باعيانها. فليكن الخطّ المعلوم خطّ بج

---

176 وليخرج: ولعقد
177 وليكن: ولكن؛ ده: اه
178 فلانّ: ولان
181 وثمنًا: وسمل

DE, as for instance if we want BG to be $1\,{}^{3}/5$ times (DE),
184 so that the ratio between them is 8 : 5.| We draw AB per-
pendicular (to BG), and divide off equal sections, making
AB 8 of those (sections) and AD 5 of them. We join AG and
185 draw DE parallel to BG.| Then DE is the required line.
Similarly if we want to find another line whose ratio to the
given line is any other given ratio, then we carry out exactly
the same operations.|

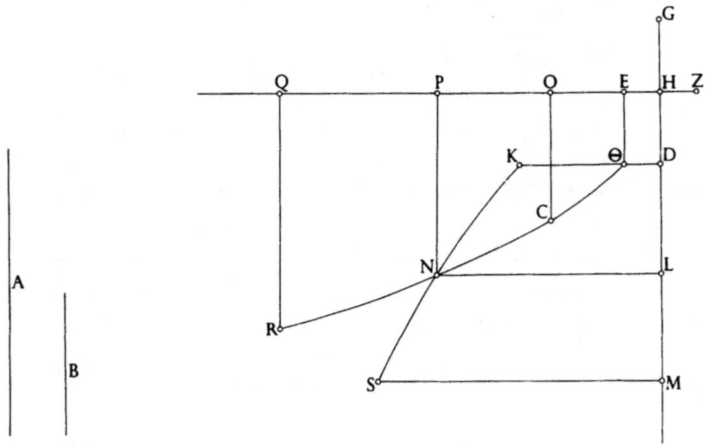

Fig. 10

186 Then after that Diocles shows how to find a cube which is twice a (given) cube by means of a geometrical proof.| Prop. 10
187 We make the given line A. Then we must seek another
188 line such that the cube of A is twice (that line's) cube.| We
construct two lines, GD, EZ, intersecting at right angles;
189 let GH, HD each equal $^1/4$ A.| We make B (equal to) $^1/2$ A;
let EH, ZH each equal $^1/4$ B. Draw θD, Eθ perpendicular
190 (to GD, EZ respectively), and make DK equal to GD.| Then
it is obvious that

$$ZE = E\theta,$$
since A = 2B.|

191 Produce GD, ZE in a straight line to points M, Q and mark

ونريد ان نجد خطّاً كخطّ ده حتى يكون خطّ بج مثلاً واجزاء
لخطّ ده كانّا اردنا بج مثله مع ثلثة اخماسه حتى تكون
184 نسبته اليه نسبة ثمنية الى خمسة. فنخرج خطّ اب على زوايا
قائمة ونفصّل خطوطاً مساوية بعضها لبعض ونجعل خطّ اب ثمنية
منها ونجعل خطّ اد خمسة منها ونصل خطّ اج ونخرج خطّ
185 ده موازيًا لخطّ بج. فيصير خطّ ده هو الخطّ المطلوب
وكذلك ايضًا ان اردنا ان نجد خطًّا آخر نسبته الى الخطّ
المعلوم نسبة اخرى معلومة ايّ نسبة كانت فانّا نعمل هذه
الاشياء باعيانها ⟨٥⟩

ي

186 ثمّ يبيّن من بعد ذلك ذيوقليس كيف نجد مكعّبًا
187 يكون ضعف مكعّب ببرهان هندسي. [٥] نجعل الخطّ المعلوم
خطّ ا وينبغي ان نطلب خطًّا آخر يكون مكعّب خطّ ا مثلي
188 مكعّبه. فنضع خطّين يتقاطعان على زوايا قائمة وهما خطّا جد
هز وليكن كلّ واحد من خطّي جح حد ربع خطّ ا.
189 ونجعل خطّ ب نصف خطّ ا وليكن كلّ واحد من خطّي
هح زح ربع خطّ ب. وليخرج خطّا طد هط على زوايا قائمة
190 ونجعل خطّ دك مساويًا لخطّ جد. وهو بيّن انّ خطّ زه يكون
191 [ايضًا] مساويًا لخطّ هط ⟨لانّ خطّ⟩ ا مثلا خطّ ب. ⟨فليخرج

---

184 زوايا: زاوبا

185 نسبته: له

on lines DM, EQ many points close to each other, as L, M, O,
192 P, Q.| Draw from points L, M on line HM perpendiculars
LN, MS, and from points O, P, Q perpendiculars to line HQ,
193 namely OC, PN, QR.| Now we made the line drawn from
D to K equal to GD. So we set D as center, and with radius
equal to GL draw a circle, and mark N at the place where
194 it cuts LN.| Similarly we draw a circle with center D and
195 radius GM, and mark S at the place where it cuts MS.| Let
us draw with the curved ruler a line passing through points
K, N, S and the other points marked in this way: that is line
196 KNS.| We operate similarly on the other line: we draw a
circle with center E and radius ZO; let it cut OC at C.|
197 Similarly we adopt the same center E, and radius ZP, ZQ
(in turn), and draw circles: let these cut PN, QR in N, R
198 (respectively).| Likewise we draw with the curved ruler a

خطًّا جد زه على استقامة الى نقطتي م ق⟩ ولنتعلّم على
خطّي دم هق نقطًا كثيرة متقاربة وهي نقط ل م ع ف⟨ق⟩،
192 وليخرج من نقطتي ل م من خطّ حم عمودا لن مس
وليخرج من نقط ع ف ق اعمدة على خطّ حق وهي خطوط
193 عش فن قر[ه]. و⟨قد جعلنا⟩ الخطّ الذي يخرج من نقطة
د الى نقطة ك مساويًّا لخطّ جد فنجعل نقطة د مركزًا وندير
ببعد مثل جل خطًّا محيطًا بدائرة ونتعلّم على موضع قطعها خطّ
194 لن نقطة ن. وكذلك ايضًا نرسم على مركز د وببعد جم
خطًّا محيطًا بدائرة ونتعلّم على موضع قطعها لخطّ مس نقطة
195 س. ولنرسم بالمسطرة التي تعطف خطًّا يمرّ بنقط ك ن س
وسائر النقط التي تتعلّم على هذه الجهة وهو خطّ كنس.
196 وكذلك ايضًا نعمل في الخطّ الآخر فنرسم على مركز ه
وببعد زع خطًّا محيطًا بدائرة وليقطع خطّ عش على نقطة ش.
197 وكذلك نضع المركز ه نفسها والبعد كلّ واحد من خطّي
زف زق ونرسم خطّين محيطين بدائرتين وليقطعا خطّي فن
198 قر[ه] على نقطتي ن ر. وكذلك نرسم بالمسطرة التي

---

191 ولنتعلّم: فلىعلم؛ كثيرة: ما كره؛ متقاربة: متقاوىه
192 مس: منس (del. ن)؛ على: ىعطه add. sed del.
193 جد: حل؛ جل: جر
197 نضع: ىصه

line passing through points θ, C, N, R, namely line θCNR.
Then lines KNS, θCNR cut one another at some point: let
199 them cut at point N.| Draw perpendicular NL to line HM and
perpendicular NP to line HQ. Then I say that

$A^3 = 2NL^3$.|

200 For $GL = DN$,

so $4LH.HD + DL^2 = DL^2 + LN^2$.|

201 We subtract DL^2, which is common, and the remainder

$4LH.HD = A.HL = LN^2$.|

202 So $A : LN = NL : LH$.

Similarly we prove that

$B.HP = PN^2 = B.LN$.|

203 So $B.LN = PN^2 = HL^2$.|

204 So $NL : LH = LH : B$.

تعطف خطاً يمرّ بنقط ط ش ن ر وهو خطّ طشنر. فخطا
ك ن س طشنر يقطع احدهما الآخر على نقطة ما فليتقاطعا
199 على نقطة ن. وليخرج عمود ن ل على خطّ ح م ⟨وعمود ن ف
على خطّ ح ق⟩. فاقول انّ مكعّب خطّ آ ضعف مكعّب خطّ
200 ن ل، وذلك انّ خطّ ج ل مساو للخطّ الذي يخرج من نقطة
د الى نقطة ن. فالذي يكون من ل ح في ح د اربع مرّات مع
المربّع الكائن من د ل مساو للمربّعين الكائنين من خطّي
201 د ل ل ن. ونسقط المربّع الكائن من خطّ د ل المشترك
فالباقي الذي هو المجتمع من ⟨ضرب⟩ ل ح في ح د اربع
مرّات الذي هو مثل المجتمع من ضرب آ في ح ل مساو
202 لمربّع خطّ ل ن. فنسبة خطّ آ الى خطّ ل ن كنسبة خطّ ن ل
الى خطّ ل ح. وكذلك نبيّن انّ المجتمع من ضرب خطّ
ب في خطّ ح ف مساو لمربّع خطّ ف ن الذي هو مثل
203 المجتمع من ضرب ب في خطّ ل ن. فضرب خطّ ب في
خطّ ل ن مساو للمربّع الكائن من خطّ ف ن الذي هو مثل
204 المربّع الكائن من خطّ ح ل. فنسبة خطّ ن ل الى خطّ ل ح

---

198 بنقط: بنقطه

199 ح م: طم

200 الكائنين: الكائن

201 نسقط: سقط: خطّ ل ن: ط ل ن

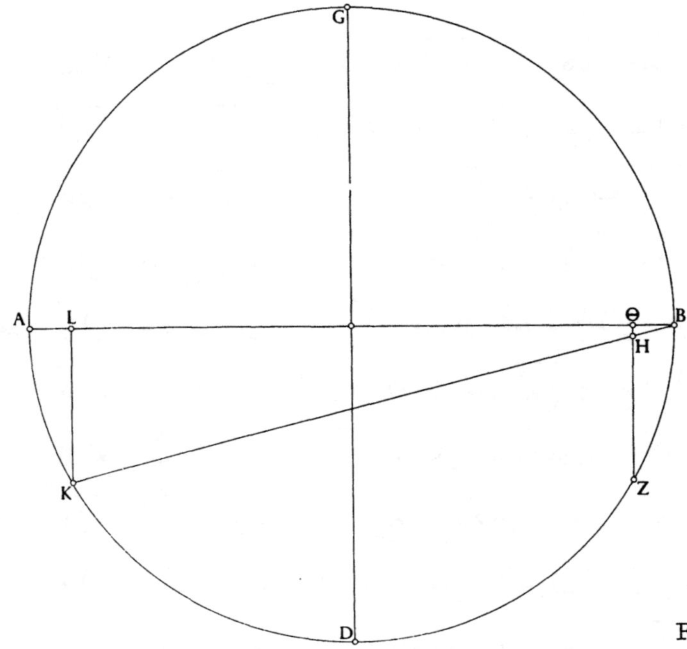

Fig. 11

And it has been shown that

NL : LH = A : NL.|

205 So between A and B are the two mean proportionals LN, LH,

so $A^3 : NL^3 = A : B.$|
206 But A = 2B,
so $A^3 = 2\ NL^3$.

So NL is the required line. We operate similarly for any other given ratio.|

207 And furthermore it is obvious that lines θCNR, KNS are parabolas, and that each of them passes through point N: hence the (conic) sections necessarily intersect each other.|

208 One may (also) find two intermediate lines by the Prop.
following method: we draw circle ABGD; let two diameters 11
in the circle, AB and GD, cut one another at right angles.|
209 We cut off two equal arcs, KD and ZD, from the circle, and
210 draw perpendicular Zθ from Z to AB. Join BK.| Then I say

كنسبة خطّ لح الى خطّ ب. وقد تبيّن انّ نسبة خطّ نل
205 الى خطّ لح كنسبة خطّ ا الى خطّ نل. ففيما بين خطّي
اب خطّا لن لح متواليان على نسبة فنسبة ⟨مكعّب خطّ ا
206 الى مكعّب خطّ نل كنسبة⟩ خطّ ا الى خطّ ب. وخطّ ا
مثلا خطّ ب فمكعّب خطّ ا مثلا مكعّب خطّ نل. فالخطّ
المطلوب هو خطّ نل. وكذلك ايضًا نعمل في النسب
207 الباقية المعلومة ايّ نسب كانت. ومن البيّن ايضًا انّ خطّي
طشنر كنس هما قطعا مخروط قائم الزاوية وانّ كلّ واحد
منهما يمرّ بنقطة ن فالقطعان يقطعان احدهما الآخر لا محالة٥

208 وقد يوجد الخطّان المتوسّطان على هذه ⟨الجهة⟩ وهي يا
ان نخطّ دائرة ابجد وليكن فيها قطران يقطع احدهما الآخر
209 على زوايا قائمة وهما اب جد، ونفصل من الدائرة قوسين
متساويتين وهما قوسا كد زد ⟨ونخرج من نقطة ز الى خطّ
210 اب عمود زط⟩ وليوصل خطّ بك. فاقول انّ خطّي زط
طب ⟨متواليان⟩ فيما بين خطّي اط طح على نسبة. والامر
في انّ نسبة خطّ اط الى خطّ طز كنسبة خطّ طز الى خطّ

---

205 خطّ ب: حط نل

206 مثلا[1]: ملا

207 طشنر: ن add. sed del.; الزاوية: الزاويا; الآخر: الاخرى

208 الآخر: الاخرى

that $Z\theta$ and θB are continuous proportionals between $A\theta$ and θH. The reason that

$A\theta : \theta Z = \theta Z : \theta B$ is clear.|

211 But I say that $Z\theta : \theta B = \theta B : \theta H$ also.
For let perpendicular KL be drawn from K to AB.|

212 Then $B\theta : \theta H = BL : LK$.
But $BL = A\theta$ and $LK = Z\theta$,

since $\widehat{ZD} = \widehat{DK}$ and $Z\theta$, KL are perpendiculars.|

213 So $B\theta : \theta H = A\theta : \theta Z$;

and, as we said, $A\theta : \theta Z = Z\theta : B\theta$.
So the four lines $A\theta$, θZ, θB, θH are in continuous proportion.|

214 Let there again be a circle ABGD, with two diameters Prop.
215 AB, GD, cutting one another at right angles.| Let us 12
cut off from the circle successive equal arcs DZ, ZH,
216 Hθ, and draw perpendiculars ZK, HL, θM to line AB.| Cut
off from the other quadrant of the circle (i. e. AD), beginning
from point D, arcs equal in size and also in number to
217 arcs DZ, ZH, θH, namely arcs DN, NS, SO.| Let the line

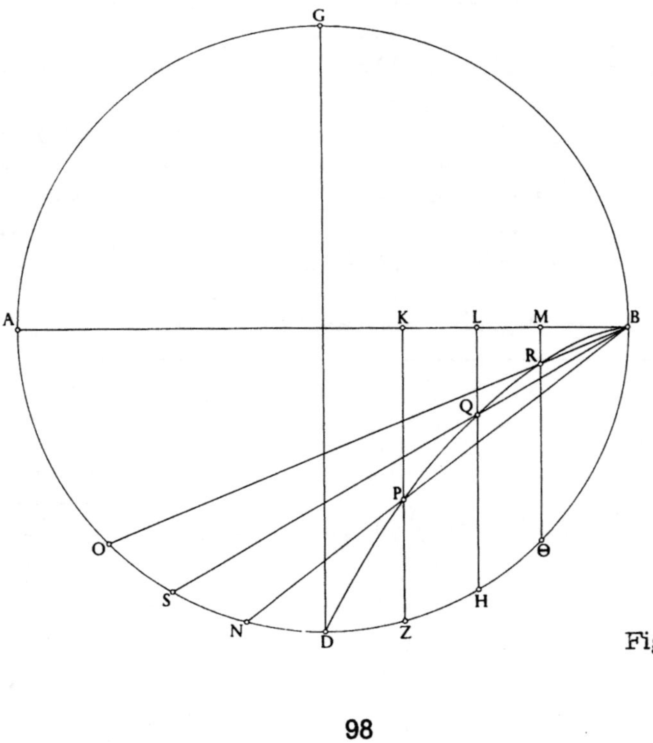

Fig. 12

211 طب بيّن. فاقول انّ نسبة خطّ زط ايضًا الى خطّ طب كنسبة خطّ طب الى خطّ طح. فليخرج من نقطة ك الى خطّ
212 اب عمود كل. فنسبة خطّ بط الى خطّ طح كنسبة خطّ بل الى خطّ لك. واما خطّ بل فهو مثل خطّ اط واما خطّ لك فهو مثل خطّ زط لانّ قوس زد مساوية لقوس دك
213 وخطّا زط كل عمودان. فنسبة بط الى طح كنسبة اط الى طز التي قلنا انّها كنسبة زط الى طب. فخطوط اط طز طب طح الاربعة متوالية على نسبة٥

214 وليكن ايضًا دائرة عليها ابجد وليكن فيها قطران احدهما قاطع الآخر على زوايا قائمة وهما قطرا اب جد. يب
215 ولنفصل من الدائرة قسيًا متوالية مساوية بعضها لبعض وهي قسي دز زح حط ولنخرج الى خطّ اب اعمدة زك حل طم.
216 ولتفصل من عند نقطة د من الرابع الآخر من الدائرة قسي مساوية لقسي دز زح طح يكون عددها مساويًا لعددها ايضًا
217 وهي قسي دن نس سع. وليقطع الخطّ الذي يصل بين نقطة ب ونقطة ن خطّ زك على نقطة ف وليقطع الخطّ الذي يصل بين نقطة ب ونقطة س خطّ حل على نقطة ق وليقطع ايضًا

---

212 واما[1]: فاما
213 طح[1]: حط ح
215 دز: از: اعمدة: اعموده

joining B to N cut ZK at P, and the line joining B to S
218 cut HL at Q, and the line joining B to O cut ΘM at R.| Then
it has been shown in the preceding proposition that ZK
and KB are continuous proportionals between AK and KP.
Similarly HL and LB are continuous proportionals between
AL and LQ, and ΘM and MB are continuous proportionals
219 between AM and MR.| So if we construct the perpendiculars
closer than those we mentioned, and mark points on them
as we marked P, Q and R, and draw through all these points
220 by means of the curved ruler line BRQPD,| then it is obvious
that if we mark on it (line BRQPD) a point (e. g.) P, and
draw perpendicular PK from it (P) to AB, the result is that
ZK and KB are continuous proportionals between AK and KP.|

221 Then since that is established, if we take line A and Prop.
the ratio of line B to line G as given, and want to find a line, 13
e. g. S, such that

$$A^3 : S^3 = B : G,|$$

222 then we draw circle DHZ, making its radius equal to A. In
it too let there be two diameters DZ, ΘH, cutting one

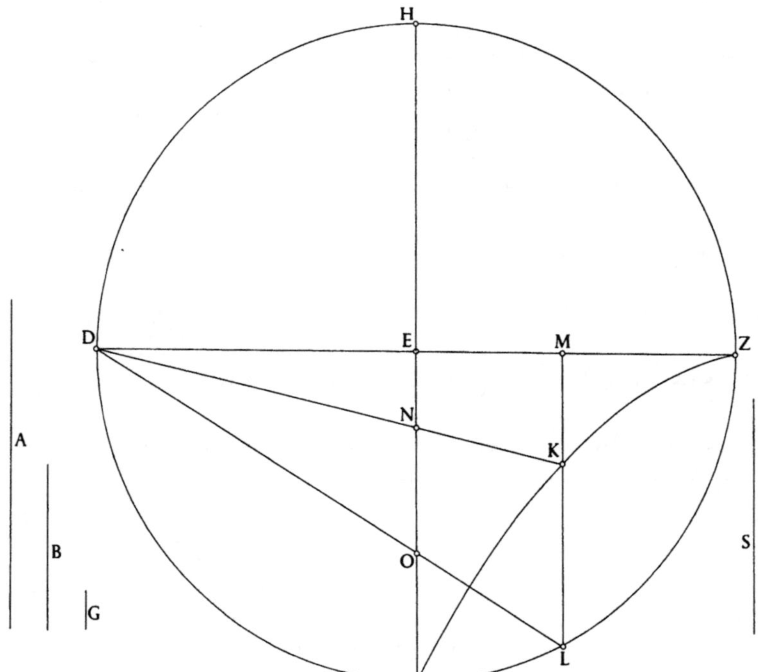

Fig. 13

الخطّ الذي يصل بين نقطة ب ونقطة ع خطّ طم على نقطة
218 ر. فقد تبيّن ممّا قلنا فيما تقدّم انّ خطّي زك كب
متواليان فيما بين خطّي اك كف على نسبة وكذلك ايضًا
يكون خطّا حل لب متواليين ‹فيما› بين خطّي ال لق
على نسبة ويكون خطّا طم مب متواليين فيما بين خطّي ام
219 مر على نسبة. فان جعلنا الاعمدة اكثر تقاربًا من التي
ذكرنا وتعلّمنا عليها نقطًا كما تعلّمنا نقط ف ق ر وخطّطنا
بالمسطرة التي تعطف على هذه النقط كلّها خطًّا وهو خطّ
220 برقف‹د›، فهو بيّن انّه ان تعلّمت عليه نقطة ف واخرج
منها عمود فك على خطّ اب صار خطّا زك كب فيما ‹بين›
خطّي اك كف متواليين على نسبة٥

221 يج فاذا تبيّن هذه الاشياء فانّا نجعل خطّ ا معلومًا ونجعل
نسبة خطّ ب الى خطّ ج معلومة ونريد ان نجد خطًّا ما مثل
خطّ س حتى تكون المكعّب الكائن من خطّ ا الى المكعّب
222 الكائن من خطّ س مثل نسبة خطّ ب الى خطّ ج، فنخطّ
دائرة دحز ونجعل نصف قطرها مساويًا لخطّ ا وليكن فيها
ايضًا قطران قاطع احدهما الآخر على زوايا قائمة وهما خطّا دز

---

218 مبتواليين: متواليان (bis)؛ لق: لح

220 كف: كح؛ متواليين: مواله

221 خطّ ا[1]: حطا ا؛ نجد: بجل

223 another at right angles.| Let a line be drawn in the way we
described, namely line ZKΘ, and let

$$DE : EN = B : G.|$$

224 Join DN and produce it until it reaches point K of the line
we described. Draw KM as perpendicular to line DZ, and
225 produce it (on the other side) to (meet the circle at) L.| Join
DL, and let it (DL) cut EΘ in O. Let S be made equal to EO.
Then I say that the required line is S.|

226 For it has been shown in the preceding propositions
that LM and MZ are continuous proportionals between DM
and MK; and when four lines are in continuous proportion,
the ratio of the first to the fourth equals the ratio of the
227 cube on the first to the cube on the second.| So

$$B : G = DE : EN = DM : MK = DM^3 : LM^3 = DE^3 : EO^3 = A^3 : S^3.|$$

228 So if we want to find another line such that the ratio of
229 its cube to the cube on A equals the ratio of B to G,| then

223 طح، وليرسم خطّ على ما وصفنا وهو زكط وليكن نسبة
224 خطّ ده الى خطّ هن كنسبة خطّ ب الى خطّ ج، وليوصل
خطّ دن وليخرج حتى ينتهي من الخطّ الذي وصفنا الى نقطة
ك. وليخرج الى خطّ دز عمود كم وليخرج الى نقطة ل،
225 وليوصل خطّ دل ⟨وليقطع خطّ هط على نقطة ع⟩ وليكن خطّ
س مساويًا لخطّ هع. فاقول انّ الخطّ المطلوب هو خطّ س،
226 لانّه قد تبيّن في الاشياء التي تقدّمت انّ خطّي لم مز
متواليان فيما بين خطّي دم مك على نسبة واذا كانت اربعة
خطوط متوالية على نسبة فانّ نسبة الاوّل منها الى الرابع
كنسبة المكعّب الكائن من الاوّل الى المكعّب الكائن من
227 الثاني، فنسبة خطّ دم الى خطّ كم التي هي مثل نسبة خطّ
ده الى خطّ هن التي هي كنسبة خطّ ب الى خطّ ج هي
كنسبة المكعّب [الى المكعّب من] الكائن من خطّ دم الى
المكعّب الكائن من خطّ لم التي هي كنسبة المكعّب
الكائن من خطّ ده الى المكعّب الكائن من خطّ هع وذلك
كنسبة المكعّب الكائن من خطّ ا الى المكعّب الكائن من
228 خطّ س. فان اردنا ان نجد خطًّا آخر حتى تكون نسبة

---

224 كم: كح

226 تقدّمت: تقدمت ا؛ مز: من

227 هن: هر

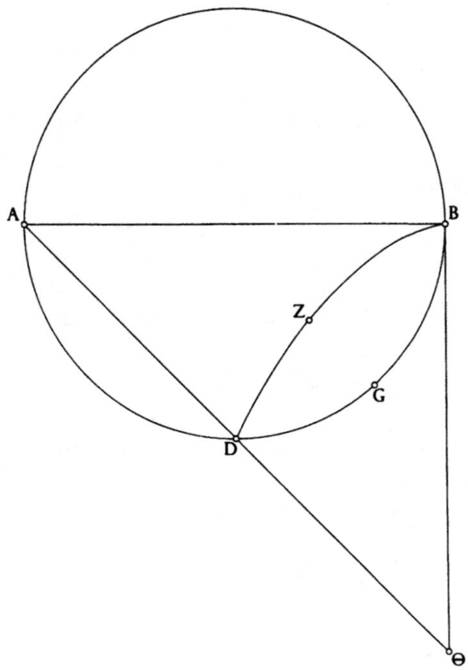

Fig. 14

it is clear that if we make the square on A equal to the
product of S and another line (X), then the required line
230 is X.| For

$$X : A = A : S,$$

$$\text{so } X^3 : A^3 = A^3 : S^3 = B : G.$$

231 Since that is so, we shall construct a triangle expressly Prop.
prepared for our needs, so that we do not have the trouble 14
of carrying out that construction for each of the given lines.
232 We draw Bθ at right angles to AB and make Bθ equal to AB.
So when we take arc AD on the semi-circle AGB, and construct the rest of the figure, we get in this diagram a right-angled triangle, ABθ, with right angle at B and AB equal
233 to Bθ,| and two lines drawn in it from its vertex B to base

الكعّب الذي يكون منه الى الكعّب الذي يكون من خطّ آ
مثل نسبة [الكعّب الذي يكون من] خطّ ب الى [الكعّب
229 الذي يكون من] خطّ ج، فهو بيّن. انّا ان جعلنا المربّع
الكائن من خطّ [ه] آ مساويًا للمجتمع من ضرب خطّ س في
230 خطّ آخر صار [ذلك] الخطّ المطلوب ذلك الخطّ الآخر، وذلك
انّه تصير نسبة ذلك الخطّ الى خطّ آ كنسبة خطّ آ ⟨الى خطّ
س فنسبة الكعّب الكائن من ذلك الخطّ الى الكعّب
الكائن من خطّ آ⟩ كنسبة الكعّب الكائن من خطّ آ الى
الكعّب ⟨الكائن من خطّ س التي هي كنسبة خطّ ب الى
خطّ⟩ ج ٥

231 واذا كان ذلك كذلك فانّا نعمل مثلّثًا عتيدًا مهيّأً لما يد
نحتاج اليه لئلّا نتكلّف عمل ذلك في كلّ واحد من الخطوط
232 المعلومة. فنخرج خطّ ب ط [ايضًا] على خطّ اب على زوايا قائمة
ونجعل خطّ ب ط مساويًا لخطّ اب. فاذا اخذنا من نصف
دائرة اجب قوس اد وركّبنا باقي الشكل صار لنا في هذه
الصورة مثلّث قائم الزاوية وهو اب ط زاوية ب فيه قائمة وخطّ اب
233 مساو لخطّ ب ط، وقد اخرج فيه من نقطة ب التي هي رأسه
الى قاعدة اط خطّا ب ز د ب ج د يلتقيان على نقطتين

---

231 المعلومة: المعموله

232 زوايا: زاويا؛ وركّبنا: وتركنا؛ الزاوية: الروايا

233 ب ز د: ب ر ح

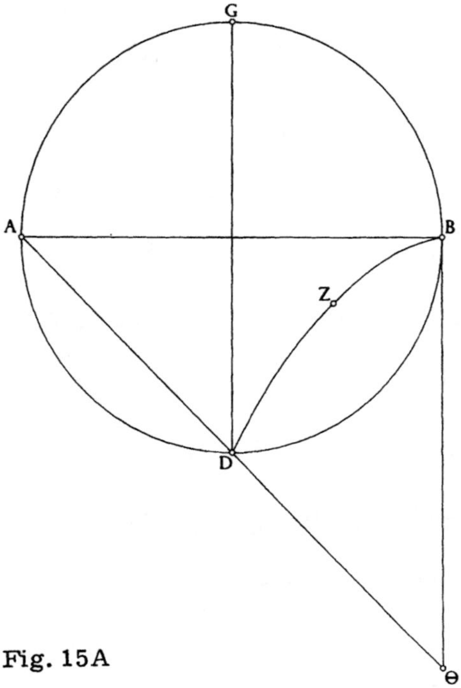

Fig. 15A

234 AΘ, BZD and BGD, which meet at two common points;| of
these BGD is a segment of the circumference of a circle,
and BZD is the line we mentioned previously. The usefulness
of that (construction) will become clear subsequently.|

235 Furthermore we make a circle ABGD in which there Prop.
are two diameters, AB and GD, perpendicular to each other. 15
Then we draw line BZD as we drew it in the preceding
diagram, and draw line ADΘ.|

236 We construct a triangle, ABG, in which AB is equal
to BG and angle B is right. We also draw two lines, BLS,
BNS, of which BNS is a segment of the circumference of
237 a circle, and BLS is the line previously mentioned.| Let
line D be given, and let the ratio E : Z be given. We make
AH equal to D, and pass through H a line HKN, parallel
to BG.|

238 Let AH : HK = E : Z.

Join AK and produce it to (meet BLS in) L. Let us pass
through L a line, MLR, parallel to BG, and connect points
239 A and R by line AR; let it cut HKN at O.| Then it is clear

234 مشتركتين لهما، وخطّ بجد منهما قطعة من خطّ محيط بدائرة وخطّ بزد الخطّ الذي تقدّم ذكره. واما منفعة ذلك فانّها تتبيّن من بعده

235 نضع ايضًا دائرة ابجد التي فيها قطرا اب جد يه
اللذان احدهما قائم على الآخر على زوايا قائمة. ثمّ نخطّ ايضًا خطّ بزد على ما وصفنا في الصورة المتقدّمة ونخرج خطّ ادط.

236 ونضع مثلّث ابج[د] الذي خطّ اب فيه مساو لخطّ بج[د] وزاوية ب فيه قائمة ونخطّ ايضًا خطّي بلس بنس اللذين خطّ بنس منهما قطعة من المحيط بالدائرة وخطّ بلس الخطّ

237 الذي تقدّم ذكره. وليكن خطّ د معلومًا ولتكن نسبة خطّ ه الى خطّ ز معلومة ونجعل خطّ اح مساويًا لخطّ د ونجيز على

238 نقطة ح خطًّا موازيًا لخطّ بج وهو خطّ حكن. لتكن نسبة خطّ اح الى خطّ حك كنسبة خطّ ه الى خطّ ز وليوصل خطّ اك وليخرج الى نقطة ل ولنجز على نقطة ل خطًّا موازيًا لخطّ بج وهو خطّ ملر ونصل بين نقطة ر ونقطة ا بخطّ ار

239 ⟨وليقطع خطّ حكن على نقطة ع⟩. فقد تبيّن من الاشياء التي تقدّمت انّ نسبة خطّ ام الى خطّ مل التي هي ⟨مثل⟩ نسبة

---

235 خطّ بزد: حط ادط برد (del. ادط)

237 خطّ ه: الى ه؛ حكن: طج

238 اح: ار؛ حك: بد؛ ولنجز: ولىحار

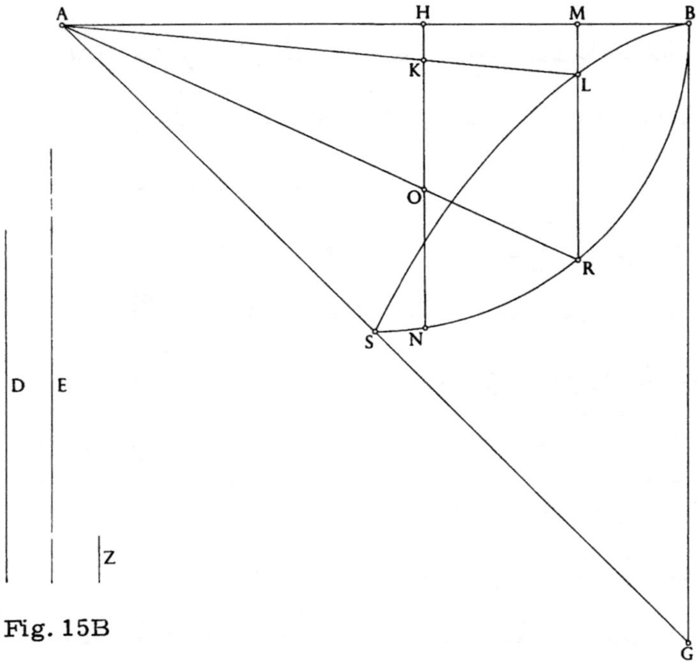

Fig. 15B

from the preceding propositions that

$$E : Z = AH : HK = AM : ML = AM^3 : MR^3 = AH^3 : HO^3.$$
And AH = D.

So the required line is HO.|

240 But if D is longer than AB, we bisect it (D), then bisect its half, and continue this procedure until we are left with
241 a line less than AB.| Then we make AH equal to the resulting line, and construct the multiple which the given line (D) is of AH, and take a line (X) such that X is that (same) multiple of HO. Then X is the required line.|

242 But (even) if points K and O are on the other side of arc BNS, there is no difference between the two cases. For the procedure is one and the same, and therefore the proof (is too). For

$$E : Z = AH^3 : HO^3.$$|

243 Similarly too for the ratio of the sphere on one of the lines to that on the other, or the cylinder to the cylinder,

خطّ اح الى خطّ حك التي هي مثل نسبة خطّ ه الى خطّ ز
كنسبة المكعّب الكائن من خطّ ام الى المكعّب الكائن
من خطّ مر التي هي مثل نسبة المكعّب الكائن من خطّ
اح الى المكعّب الكائن من خطّ حع وخطّ اح مساو لخطّ
240 د فالخطّ المطلوب هو خطّ حع. فان كان خطّ د اطول من
خطّ اب قسمناه بنصفين ثمّ قسمنا نصفه بنصفين ولا نزال
241 بعمل ذلك حتى يبقى لنا خطّ ما اقصر من خطّ اب. فنجعل
الذي يبقى لنا خطّ اح وننظّم اضعاف الخطّ المعلوم بخطّ
اح ونأخذ خطًّا ما يكون اضعافه بخطّ حع مثل تلك الاضعاف
242 فيكون ذلك الخطّ هو الخطّ المطلوب. وان كانت نقطتا
ك ع في الناحية الاخرى عن قوس بنس لم يكن بين
الامرين فرق وذلك انّ العمل عمل واحد نفسه ولذلك
البرهان لانّه تكون نسبة خطّ ه الى خطّ ز كنسبة المكعّب
الكائن من خطّ اح الى المكعّب الكائن من خطّ حع.
243 وكذلك ايضًا نسبة الكرة التي تعمل على احد الخطّين الى
التي تعمل على الآخر والاسطوانة الى الاسطوانة او شيء ممّا
تملأ به الكرة او الاسطوانة او ايّما يتبع جوهرهما فانّ العمل

---

239 اح[1]: دح؛ اح[3]: ان

241 وننظّم: وسطرم

242 اح: ام؛ حع: مع

243 ايّما: اما؛ جوهرهما: حوهرهواما

or any of the (?)subdivisions of the sphere or cylinder, or whatever follows their essential nature: in every case the procedure and the proof is the same.|

244 From these propositions it has become obvious how Prop. 16
we may find two lines intermediate between two given lines such that the four are in continuous proportion. The method is as follows: we make A and B the two lines between which we want to find two (other) lines so that the four are
245 in continuous proportion.| Let any other line whatever, G, be constructed. Set

$$A : B = G^3 : D^3,$$

as we have shown above, and let there be constructed a third line, E, in continuous proportion with G and D, and a third line, Z, in continuous proportion with D and E.|
246 Then G, D, E, Z are in continuous proportion. So

$$G : Z = G^3 : D^3 = A : B.|$$

247 Construct a line H such that

$$A : H = G : D,$$

and another line θ such that

$$H : \theta = D : E.$$
$$\text{Then } A : \theta = G : E.|$$

248 And G, D, E, Z are in continuous proportion,

$$A : B = G : Z,$$
$$\text{and } A : \theta = G : E.|$$

والبرهان في جميع ذلك واحده

يو

244 قد تبيّن من هذه الاشياء كيف نجد خطّين بين
خطّين معلومين حتى تتوالى الاربعة [الخطّ] على نسبة وذلك
انّا نجعل الخطّين المعلومين اللذين نريد ان نجد بينهما
245 خطّين لتتوالى الاربعة على نسبة خطّي آ ب، وليوضع خطّ ما
ايّ خطّ كان وهو خطّ ج ولتكن نسبة خطّ آ الى ب
كنسبة المكعّب الكائن من ج الى المكعّب الكائن من د
كما بيّنّا فيما تقدّم وليوجد لخطّي ج د خطّ ثالث مناسب
لهما وهو خطّ ه ولخطّي د ه خطّ ثالث مناسب لهما وهو
246 خطّ ز. فخطوط ج د ه ز متوالية على نسبة فنسبة خطّ ج الى
خطّ ز كنسبة المكعّب الكائن من خطّ ج الى المكعّب
الكائن من خطّ د التي هي ⟨مثل⟩ نسبة خطّ آ الى خطّ ب.
247 ⟨فليوجد خطّ ما تكون نسبة خطّ آ اليه كنسبة خطّ ج الى خطّ
د وهو خطّ ح وليوجد خطّ آخر تكون نسبة خطّ ح اليه كنسبة
خطّ د الى خطّ ه وهو خطّ ط.⟩ فتصير نسبة خطّ آ الى خطّ ط
248 كنسبة خطّ ج الى خطّ ه. وخطوط ج د ه ⟨ز⟩ متوالية على نسبة
ونسبة خطّ آ الى خطّ ب مثل نسبة خطّ ج الى خطّ ز ونسبة

---

244 خطّي آ ب: واحده خطي آ ب الخطّين (واحده et الخطّين del.)

247 خطّ ط[2]: حط ح

248 مثل: هي (bis)

249 So $\theta : B = E : Z = D : E = H : \theta$.
250 And $A : H = G : D = D : E = H : \theta$.|

251 So lines A, H, θ, B are in continuous proportion, so we have found two lines, H and θ, between lines A and B in continuous proportion with them.|

252 That is the end of what was found of Diocles' Book on Burning Mirrors. It is finished with the aid of God on high. Praise be to God, lord of the worlds, and may God bless our prince Muḥammad and his family, the good and pure, and grant full salvation.|

249 خطّ ا الى خطّ ط مثل نسبة خطّ ج الى خطّ ه. فتكون نسبة
خطّ ط الى خطّ ب كنسبة خطّ ه الى خطّ ز التي هي مثل
نسبة خطّ د[ا] الى خطّ ه التي هي مثل نسبة خطّ ح الى خطّ
250 ط. ⟨ونسبة خطّ ا الى خطّ ح كنسبة خطّ ج الى خطّ د التي
هي مثل نسبة خطّ د الى خطّ ه التي هي مثل نسبة خطّ ح
251 الى خطّ ط⟩. فخطوط ا ح ط ب متوالية على نسبة فقد وجدنا
فيما بين خطّي ا ب خطّين متواليين معهما على نسبة وهما
خطّا ح ط.

252 هذا آخر ما وجد من كتاب ذيوقليس في المرايا المحرقة
تمّت بعون الله تعالى والحمد لله ربّ العالمين وصلى الله
على سيّدنا محمّد وآله الطيّبين الطاهرين وسلّم تسليمًا كثيرًا.

Colophon of manuscript (no.393 p.1)

١٠٦

الجميع ولا يحصل من حركه نقطه ث عند نقطه ه التي هي مركز اي اصل
سا الازمان المساويه ز وانما متساويه لانها تقرب منها تارة وتبعد اخرى
ويحصل ها الطبع لنقطه ث عند نقطه ث فيحصل الزوايا عند نقطه
ث في الازمان المساويه متساويه وذلك ما اردنا ان نبين ٥

بسم الله الرحمن الرحيم اللهم امن
كتاب ذيوقليس في المرايا المحرقه قال ذيوقليس نوفيون
المهندس الذي من اهل ارقاديا سال فيثون مسائله منها كيف
نجد بسيط مراه متى وضع مقابل الشمس اجتمعت الشعاعات التي تنعطف
منه الى خط محيط دائره   واما ابيدوروس المنجم فانه لما صار الى
ارقاديا وقدم منها سالنا كيف نجد بسيط مراه متى وضع مقابله

١٠٧

الشمس اجتمعت الشعاعات التي ينعطف منها الى نقطة ما حدث ما يحكى بانا
نروم ان نبين الجواب فيما سال عنه فوثون وما سال عنه ابيوذورس
ونستعمل في ذلك المقدمات التي قد بيناها لكي نعلمها واحدى ما تين
المسئلتين وهي التي يطلب فيها عمل مراة تجمع شعاعها الى نقطة ان صح
كان الذي طلبها دوسيثاوس واما المسئلة الاخرى فانها لما كانت علما فقط
ولم يكن لها عمل يستحق ان نشهره لم نعمل وقد بينا بالبيت براهين كل واحدة
من ما بين المسائل واوضحناها   وبسيط المراة المحرقة الذي
يجمع اليه هو البسيط الذي يحيط به الشكل الذي يحدث من قطع المخروط
القائم الزاوية اذا ادير حول الخط الذي يقسمه بنصفين ما يعرض لهذا
البسيط ان ينعطف جميع شعاعاته الى نقطة واحدة وهي النقطة التي تبعد
بنصفين ما يعرض لهذا البسيط ان ينعطف جميع   بعد ما من ذلك البسيط
مثل ربع الخط الذي يقوم عليه الاعمدة التي تخرج الى السهم وكلما زيد
في ذلك البسيط مثل ربع الخط الذي يقوم عليه الخط زيادة معلومة مثل قطعة
دائرة تزاد في قطع المخروط الذي ذكرنا فان الشعاعات التي تنعطف
من تلك الزيادة تنعطف ايضا الى تلك النقطة بعينها متبددة في قوس الخارج
التي حول النقطة وهي تحدث الاحراق اقوى من حدوث الاحراق في التي
تكون من بسيط الكرة وذلك لان انعطاف الشعاعات من بسيط الكرة
انما يصير الى خط مستقيم لا الى نقطة وان كان قد ظن قوم انها تصير الى
المركز والشعاعات التي تجتمع في هذا البسيط الى موضع واحد انما تنعطف
من بسيط قطعة كرة اقل من نصف الكرة وان كانت المراة من نصف كرة او
من اكثر من نصفها لم ينعطف الى ذلك الموضع غير تلك الشعاعات التي انقطعت
من اقل من نصف الكرة والمسئلة التي سال عنها نحن نبين نعمل ايضا بقطع
المخروط القائم الزاوية اذا ادير حوله ما من الادارة ونبين ذلك
فيما بعد وقد احتيج الى عمل مراة محرقة من غير ان تطلب بازاء الشمس
وهي ما بين موضع واحد بعينه وبين ساعات النهار من غير مقياس وذلك
انها تحرق ابدا ما ينعطف اليه الشعاعات واعظمها يكون ابدا الى موضع سائر
المطلوب وهذا العمل شيء يحتاج ذكره الاطلاع فيه الى من يعلم عمل المراة

١٠٨

لكنه يحدث عن الشكل الذي ذكرنا بسيطة وكمن واكثرون اولا اعتقاد ح
يستعملها المنجمون وهي ان كل بسيطة يستعمل من البسيطة التي على الارض فانها تقوم
مقام مركز الارض وقد تسحك قوم ويشنعون على اصحاب العلوم التعليمية
ويقولون انهم يبنون امرهم على اساس ضعيف ويزعم بعضهم انه قد علم
بمقادير اقطار الاكر والة وا حد بعضها ويزيلوا حد منها اكثر من الاخر
ماكثر من بضع الف بسطاذون وزعم بعضهم ان ذلك اكثر من خمسين
الف بسطاذيون ومال الناس الى هذا القول الثاني اكثر لانهم ونقدا فيه
ماقاله القدماء ويقولون انه ان وجد السبيل الى ان لا يستعمل فيها هذا
الا فضل ولم يضطر فاليه الحاجة في امر الالات الساعات التي يستعمل فيها
الظل فالاصلح ان لا يستعمل ركن عن ذلك وزعمنا الموضع حتى [illegible] ماكانوا
فيه يعدلون ان البسيطة التي ذكرنا تقوم مقام مركز الارض في
والعالم فيقع في ذلك القياس الواجب الذي يستعمله في هذا الموضع
ونحوه ويتبين ان الذي يعرض في امر القياس شبيه بما كان يعرض فيها
لو وضعت على الكمية في المركز فان الالات الشعاعات التي يستعمل
فيها الظل فاكانت بنيها ذا الا على الشعاعات من غير قياس يكون بها يبلغ
في الاستقصاء والكمية حتى لا يكون في ذلك دون غيره وقد يمكننا نحن
ان نصطر انضاد في عمل [illegible] الالات ان احسنت ذلك لان فيها شيء
يستحق ان يتعجب منه وذلك انه يمكننا ان نعمل ما قاله في ذلك غيرنا وهل
اصابوا ام لا فيما ادعوا في ذلك اولم يصيبوا ولانه مع ذلك بالجملة لا يذهب
في هذا العمل شيء يحتاج اليه في الالات الساعات التي ذكرناها وانا ارى
ان الذي ينبغي ان نبينه ههنا هو علوا [illegible] والحال انه ما يستحق
ان نخطر فيه وتتعلم ما امر المقاييس التي يستعملها المنجمون فانها
مستقيمة صحيحة اذا عملت على الرسوم القديمة التي كان يعمل عليها
الالات الساعات التي يستعمل فيها الظل ويمكن يكسر امر البسيط الذي
يستعمل فيها لا يمكن ان يعمل ذكرها فها نحن نعملها هذا او ينبغي ان نعلم
بالجملة انه دون ذلك [illegible] وما [illegible] كان [illegible] ما
ذلك ليكون يصدق ما قلنا وقد يمكن ان نعمل من اساس من المنزا يا

المرآه يكون نقط كل واحده منهما دراهمين فيكون احداهما بقوله ما خط
محيط بزاويه والاخرى على قطع المخروط القائم الزاويه حتى يمكنك ان تعمل
من كل واحده منهما والاوراق بسبع حلقا حسنها سعلم تحصل افرادها
اوراقا وتماس اوراق واحده منها الاوراق الاخرى امر يستحق "
ان يشاهد وذلك لبسيطة المراه التي نظرتها مقدار قدم او سبح اشبر
كان تحرق جميع ما بد قوامها موضع الاحراق عن الكسب من اجل ان تحرق
بسهوله اذا كان الخط سعه اعمال وذلك لانه اذا صوعف موضع الاحراق
سبع مرات لم يمكن الا يكون الفصل بينهما بين الاول فصلا كبيرا جدا
ولكن يقول انه يمكن ان يعمل المرآه من زجاج حتى يعرض فيما شيء حاد هوائه
سيدي منها مصابيح توجد بها نار من المصباح وبعد النهار والقرابين حتى
قال ان يفتح النار ترى النهار كما يلقا انه يكون في بعض المدن التي انساسه
عما وخاصه في الايام التي تعظمها ينتحب اهل تلك المدن في ذلك وهذا
شيء نعمله نحن اعمالا في قطع المخروط القائم الزاويه شكل م والخط
الاول نقسمه بنصفين خط ا ر ونصف الخط الذي يقول عليه الاعمدة خط
ح ح الذي خط ت ه والخط الذي نقسمه مساويا لخط س ح و
لنقسم خط ر بنصفين على نقطة د ولنخرج خطا مماس القطع ليس لا يقع
وهو خط ط ا ولنخرج خطا ط ح عمود اعلى ا د فقد علمنا ان خط ا ت
مساو لخط وان الخط الذي يخرج من نقطة ك ي خط ط ا على زوايا
قائمه على خط ا د خارج بنقطة ه فلنخرج خط ر ك ي خط ط ا
على زوايا قائمه ولنوصل خط د ك يكون ح د مساويا لخط ح
وخط ح ت مساويا لخط ت ه فخط ح د مساويا لخط ه و سقط
خط ح ه المشترك فيبقى خط ح ت مساويا لخط ه د وخط ح ك
مساو لخط ا د لخط ا ت مساو لخط ه د وخط ت د مساو
لخط د ه وذلك ان خط ت ه بنصفين على نقطة د فجميع خط
ا د مساو لخط د ر ولان مثلث ا ط ك قائم الزاويه وقد قسم
ما عليه ا ر بنصفين على نقطة د صار خطوط ا د د ط د ل
مساويه فزاويه ع مساويه لزاويه ح وزاويه ا مساويه لزاويه

ف ق ه فليجز على نقطة ك خط موازٍ لخط ا د وهو خط ط ك س ه
فزاوية ع ك ف مساوية لزاوية د ا ف المساوية لها وزاوية ع ك ف مساوية لزاوية
ف ك خ فزاوية خ ك ف مساوية لزاوية د ر ف وزاوية ف ق ه خ القائمة
مساوية لزاوية ا ش القائمة فزاوية ث القائمة مساوية لزاوية ف ق ه
القائمة فخط ك ط اذا لقي خط ا ط انعطف الى نقطة د فاحدث
فيما بينه وبين الخط ا ط زاويتين ف ق ه ر ث المتساويتين فقد تبين
انه ان اخرج من نقطة ما من النقط التي على خط ع ف م خط مماس
للقطع ووصل الخط الذي بين نقطة المماسة ونقطة د مثل خط ط د
واخرج خط ك ط موازٍ لخط ا د فان خط ك ط اذا انعطف الى نقطة
د فان الخط الذي يمر بنقطة ك المماس للقطع يكون انعطافه على زوايا
متساوية وجميع الخطوط الموازية التي تخرج من جميع النقط التي على خط
ك ع م يكون حالها مثل هذه الحال فاذا حدثت مع الخطوط المماسة زوايا
متساوية صارت الى نقطة د فان اثبت خط ا د وادير خط ك ع م
حتى يعود الى الموضع الذي منه ابتدأ وعمل سطح مقعر صقيل على السطح
الذي مر به خط ك ع م ووضع مقابله الشمس فلتنعطف شعاعات
الشمس البسيط المقعر انعطفت الى نقطة د لانها متوازية وكلما كان البسيط
اعظم كانت الشعاعات التي تنعطف الى نقطة د اكثر فان فصلنا خط
ب ط ث وجعلناه مساويا لخط ع م ووصلنا خط ث م الذي
يوتر قطعة ث ع م واتممنا عمل خط ا ر ع سطحا مائلا على السطح
الموضوع مكان خط ث م مائلا على السطح على زوايا قائمة واثبتنا
خط ث م ثم ادرنا قطع ع ث م حتى يعود الى الموضع الذي منه
ابتدأ وعملنا سطحا مصقولا شبيها بالسطح المقعر الذي وصفنا
قبالة الشمس انعطفت الشعاعات من جميع سطحه الى المحيط بدائرة
السطح القائم على ا ر ود كمال نقطة د اذا ادير مع قطع ع ث م
انعطفت الشعاعات الى الخط المحيط بالمواضع التي تحدث عند دوران
مثل النقطة وتصير الخط الذي يخرج من المركز الى الخط المحيط بتلك المواضع
مساويا لخط د ه فان اردنا ان تكون المواضع التي بمقدار اردنا جعلنا

خط د ه مساويا للخط الذي يخرج من مركز ا الى الخط المحيط بها و
اخرجنا خط ت ه على خط ب م على زوايا قائمة وعملنا مثل
الاشياء القائمة التي عملناها وينبغي ان نعلم انه لا ينبغي ان يستعمل السطح الذي
ذكرناه كله حتى لا يكون خط ب ه ت ف الشمس فيكون سائرا الينا وان
استعملنا نصف السطح الذي ذكرناه كان الذي يحدث تقطيع السطح سطح
قائم بخط ب ه كان مقاربا ولم يقع الاجزاء في جميع الخط المحيط
بالدوائر التي ذكرناها بل على صفة ما ان المنا سطحا من خط ب ح ويكون خط
ا ت قائما عليه على زوايا قائمة ويكون قائما على سطح
الذي من خط ب ت م على زوايا قائمة
ونسميه دائرة ا وهي دائرة ب ك
د ح فيكون قطعة ت على الخط المحيط
بدائرة ب ك د ونعمل قطعة
ب م من القطع على حالها ويكون خط
ب ك قائما على السطح الذي
ذكرنا مع زوايا قائمة في جميع دوران القطعة حدث من ذلك سطحا و
رسمت نقطة د دائرة ع فاسقط الشعاع على الخط المحيط بها وليصير
بعد عمود ذلك السطح في الدوائر التي نريد ان يسقط الشعاع على الخط
المحيط بها مثل خط ب د وليس يمكن ان يكون الا القطعة الى خط
محيط دائرة فقط بل بعد يكون حتى يكون انتها الخط اجزاء خط ازدياد وذلك
وذلك انه قد اخرج من نقطة ت من قطع ك ب م خط ب ح على زوايا قائمة
فيكون في السطح الذي يمر بخط ب ح خط ك ه وهو خط ب د بالمقدار
الذي نريد ان يكون الخط الذي يسقط اليه الشعاعات والدائرة على خط ب ك د فقطع
ب م من القطع المكافئ تخط على نقطة ب من خط ب ك د والمخرج وزوايا مع
زوايا قائمة على السطح الموضوع الذي هو السطح الذي يمر بخط ب ح ولنعلم
فيه خط ر ت المماس لخط ب ك د كما في الوضع الاول فهو يبين ان نقطة
د يحدث في السطح المواز للسطح الذي يمر بخط ب ح خطا ما وما
لخط ب ح خطا مساويا له ب د نسميها الثلاث خط ت ام محيط

١١٢

موازي ذكره ه بنقطة ت ويلقى الخط المحيط بها خط ما وهو خط ا ت
ولنخرج من نقطتي ب ه خطين كانا خطان موازيان لخطات ومماسا
ل ت ح ه ونصل خط ا ك ه ح فيكون زاوية ب ك م مساوية
لزاوية ب ح ع وايضا فزاوية ح ك ه مساوية لزاوية ت ك م كانت
زاوية ه ك ت الباقية مساوية لزاوية ب ك ت الباقية ولكن زاوية ت ك ه
مساوية لزاوية ك ب ه المبادلة لها فزاوية ك ب ه مساوية لزاوية
ه ك ت فخط ب ه مساو لخط ك ه ولكن خط ك ه اطول من خط ه ا و
ذلك ان خط ه ا اقصر الخطوط التي تخرج من نقطة ه الى الخط المحيط بالدائرة
فخط ب ه اطول من خط ه ا وهو ايضا اطول من خط ه ح لان خط ك ه
اطول من خط ه ح فزاوية ح ك اعظم من زاوية ك والزاوية المبادلة
لزاوية ك هي زاوية ت فزاوية ح ت اعظم من زاوية ت فجميع زاوية
ثخه ت مساوية لزاوية بخ ق فزاوية ثخه اعظم من زاوية ق والذي زاوية
قخع مساوية لزاوية ثخه فالزاوية الباقية التي هي … مساوية لزاوية
ق الباقية وزاوية ق مساوية لزاوية ك فزاوية ك مساوية لزاوية ك
فخط ب ح اعظم من خط ح ا فلنجعل خطات … على نقطة ح فقد
تبين انه اذا خرج الى الخط المحيط بالدائرة خطوط متوازية كم كانت والتقت
من خط … المحيط بالدائرة واحدثت زوايا مساوية … بين
نقطتي ا ح … خط … ما كان من
الخطوط المتوازية لخطات الخطات فعدت … فان انعطافه يكون الى نقطة
ح اقرب وما كان منها ابعد فان انعطافه
يكون الى نقطة ا اقرب ه لذلك
خطوط ت اصغر محيط دائرة … ه
المسطح الموضوع ولتكن مركزها
نقطة ت وليكن الواصل خط ب ا
وليكن خط ب ا مماسا على نقطة ح
وليكن خط د ت … لخطات على … فانه
فيكون كل واحد من قوسي د ا ا ث سدس الخط

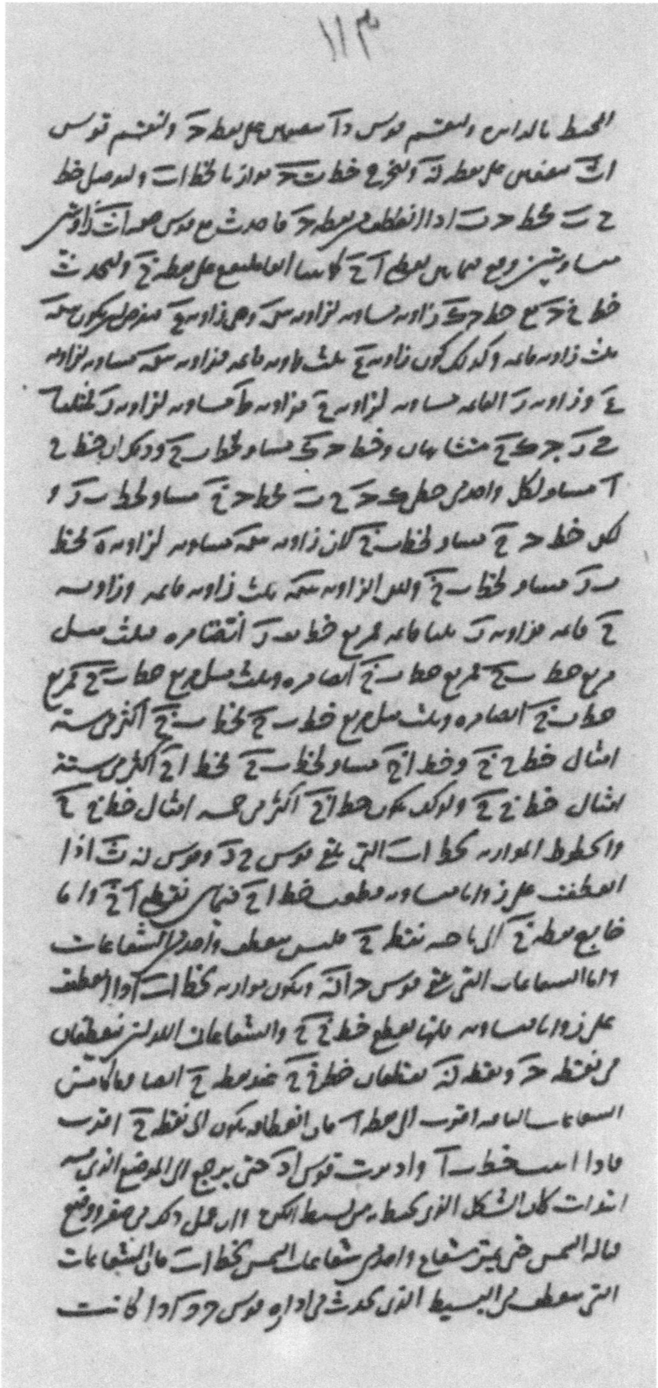

١١٣

المحيط بالدائرة ولنقسم قوس د آ بنصفين على نقطة ح ولنقسم قوس
ا ح بنصفين على نقطة ك ولنخرج خط ب ح موازيا لخط ا ت ولنوصل خط
ح ب لخط ح ب اذا انعطف من نقطة ح فاحدث مع قوس ضعفات زاويتين
مساويتين ويقع مماس لنقطة آ ح لكونها انعطف على نقطة ح وليحدث
خط خ ح مع خط ح ك زاوية مساوية لزاوية ب ح وهي زاوية ع منفصل يكون
مثلث زاوية قائمة وكذلك يكون زاوية ع مثلث زاوية قائمة فزاوية ب مساوية لزاوية
ع وزاوية ر القائمة مساوية لزاوية ع فزاوية ط مساوية لزاوية ر فمثلثا
ح ر ج ح ك ح متشابهان وخط ح ك مساو لخط ب ح وذلك ان خط ع
آ مساو لكل واحد من خطي ح ح ع ب لخط ح خ مساو لخط ب ر و
لكن خط ح ح مساو لخط ب ع لان زاوية ب مساوية لزاوية ه لخط
ب ر مساو لخط ب خ ولكن الزاوية ب مثلث زاوية قائمة وزاوية
ع قائمة فزاوية ب مثلثا قائمة فمربع خط ع ر انقص من مثلث مثل
مربع خط ب ح فمربع خط ب خ انقص من ثلث مثل مربع خط ب ح فمربع
خط ب خ انقص من ثلث مثل مربع خط ب ح لخط ب خ اكثر من ستة
امثال خط ع خ وخط ا ح مساو لخط ب ع لخط ا ح اكثر من ستة
امثال خط خ ع ولو كان يكون خط ا ح اكثر من خمسة امثال خط خ ع
والخطوط الموازية لخط ا ب التي تقع قوس ع د وقوس له ب اذا
انعطفت على زوايا متساوية وقطعت خط ا ح فيما بين نقطتي آ ع وانما
تقع نقطة خ الى ناحية نقطة ع فليس يقطع واحد من الشعاعات
وانما الشعاعات التي تقع قوس ح ا ك ويكون موازية لخط ا ب اذا انعطفت
على زوايا متساوية فانها تقطع خط ع ح والشعاعان اللذان ينعطفان
من نقطة ح ونقطة ك ينعطفان خط ع ح عند نقطة ح الى ما كان من
الشعاعات القائمة اقرب الى خط ا فان انعطافه يكون الى نقطة ع اقرب
فاذا ادير خط ب آ وادير قوس ا د حتى يرجع الى الموضع الذي منه
ابتدأت كان الشكل الذي يحيط به من بسيط المخروط الا على ذكر من صغر ومنبع
حالة الشمس حتى يقع شعاع واحد من شعاعات الشمس بخطوط فان الشعاعات
التي تنعطف من البسيط الذي يحدث من ادارة قوس ح د كلها كانت

١١٤

ينعطف على زوايا حادة فانها تنفذ الى خط ا ح واما الشعاعات
التي تنعطف من جميع البسيط الحادث من قوس ح ز ا د فانها تنفذ الى خط
ح ع والشعاعات التي تخرج من المواضع العظيمة التي لم ينعطف [illegible]
ذلك البسيط تقع عند نقطة ح والشعاعات التي هي اقرب من نقطة
آ تكون [illegible] اذا انعطفت الى نقطة ح اقرب فيكون فيما بين نقطتي
ح ح [illegible] ويكون الى نقطة ح اقرب منها الى نقطة ح فليس
ينتفع بالبسيط الحادث من قوس ح ز د [illegible] لا يقع منه شعاع
[illegible] على الموضع الذي يكون منه الاحراق ولذلك انما ينتفع [illegible] اردنا ان
نعمل مرآة محرقة من قطعة لكن ان نعمل البسيط
الذي يحدث من قوس ح ا فقط

د كيف نعمل تقعير المرآة المعرفة اذا
اردنا ان يكون القطعة التي بعينها
يكون الاحراق على بعد معلوم من
وسط بسيط المرآة وصفا خطا على بعد
معلوم على بسيطه مساويا للبعد الذي نريد وهو خط
ا ب وصيرنا خط ا د ضعف خط ب آ وجعلناه [illegible] على زوايا قائمة
ونجعله [illegible] وما يخط ا ب ونصل ه ك ونعمل خط ا ث مساويا لخط
ب ه ونصل خط ه ث فيصير ا ه ث [illegible] وايضا [illegible] ه ث
يكون [illegible] وما لخط ث ك ونعمل على خط ب آ نقطتي ح د ونعمل
خط ه ز مساويا لخط ب ح وخط ب ه [illegible] لخط ب د ونخرجها
[illegible] الجهتين [illegible] ه ث على نقطتي م ك واذا جعلنا نقطة آ
مركزا وادرنا ببعد ح م دائرة فقطعت خط ح م فلنقطع على نقطة
ك ثم [illegible] على هذا المثال حتى نقطع على نقطة ص ع وايضا فانا جعلنا ه
نقطة آ مركزا وادرنا دائرة ببعد مثل د ك فقطعت خط د ك فلنقطع
على نقطة ك ثم [illegible] والمركز على نقطة آ حتى نقطة ايضا على الخط [illegible]
[illegible] ا ح على استقامة خط [illegible] ونعمله [illegible] فيكون نقطة ك [illegible]
[illegible] ح مع قطع المخروط القائم الزاوية وذلك انا نخرج خط [illegible] الى النقطة

١١٥

رو لىس خط سه مساويا لخط ات ولنخرج خط دسه ولكن ما دا
لخط كا ولىس على خطاب على زوايا قايمه ولنوصل خط سه ك
ونخرج من نقطه ل ىم طه ك ال خط دسه ا عمله م شه ىع طف
فخط كه اذا اخرج على استقامه وقع على نقطه ى فيكون خط فه ك
مساويا لخط لد و خط م شه مساو لخط م ح و ذلك ان خط كه ك
هو نظير مربع اسه لخط لد مساو لخط نا و خط م ح مساو لخط
طا و خط ل ف الصاما
لخط لد و خط م شه ما و لخط
م ح لخط ار مساو لخط ل فه
الذى هو مثل خط ىع و خط
اط مساو لخط م شه الذى هو
مثل خط ف و خط اى مساو
لخط ى سه و هذا خط ار ه
منصاس على نقطه ت واذا الحال
ذكر لولا ان خط سه طه ك
على قطع المخروط السابع النوازه
كما بين فيما بعد ولا كرا اعضا
فقط مره ضعه ع حاد الفعليا
على خط ات لقطا كمال و ان رحنا
منها خطوطا موازيه لخط اد
وجعلنا و النقط الاخر و عطفنا
على النقط التى تحدث خطوطا من خروف و اعنى تا حتى لا تمكن لا ىحرى
وخططنا عليها خطا و عطفنا على ذكره الخط المعوج و جعلنا الشعب الشكل
الذى جرد ع عمله على هذه الحالة فان الاعراف يكون فى البسيط على
نقطه آ و ىس ذكره من الشكل الاول ٥ ٥ ولكن فقط
مربع اسه خط دك ولنضم خط ار منصاس على نقطه ت ولنعلم نقطه
ما فيما بين نقطتى ا ت وهى نقطه ح ولنجار على نقطتى سكر هطا ن

Manuscript p.116 §§ 113-124

116

موازيان لخط اد وعلى ب ه حم وعلى نقطة ه خط مواز
لخط اد وهو خط ه ث لخط م ح يتصل على خط ب ه كخط م ن
من اجل ما قدمنا في الشكل الذي قبل هذا يصير اه مربعا وخط ه ث
مساو لخط ب آ وخط ه ن مساو لكل واحد من خطي ن م ب ح فخط
م ن مساو لخط ب ح فخط م ح يتصل على خط ا ب فخط ب ح لكن
خط ب ق مساو لخط ح ث ولذلك يصير خط ق آ مساويا لخط م ح
فالدائرة التي تعمل على مركز آ وبعد مساو لخط م ح من نقطة ق فاقول
انها تقطع خط م ح فيما بين نقطتي م ح وذلك انها ان مرت بنقطة م او وقعت
دون نقطة م صار الخط الذي يخرج من مركز ا الى الخط المحيط بها اطول
من خط ح م لان زاوية ح قائمة وذلك غير ممكن وذلك انا كنا جعلناه مساويا
له فالخط المحيط بالدائرة التي ذكرنا يقطع خط م ح فيما بين نقطتي
م ح فلتقطعه على نقطة خ ولنوصل بين نقطة خ وبين نقطة آ بخط
خ ا فخط ا خ مساو لخط م ح وقد بينا ان خط ا ق ايضا مساو
لخط م ح فخط خ آ مساو لخط ا ق ولان خط ق ب مساو لخط ب ح
صار ما يجتمع من ضرب خط ا ب في خط ب ح اربع مرات مع المربع الكائن
من خط ح ا مساو للمربع الكائن من خط ا ق والمربع الكائن من خط ا ق مساو
للمربع الكائن من خط ا خ والمربعان الكائنان من خطي خ ح ح آ للمربع الكائن
من خط ا خ لان زاوية ح قائمة فالمجتمع من ضرب خط ا ب في خط ب ح اربع
مرات مع المربع الكائن من خط ح ا مساو للمربعين الكائنين من خ ح ح آ
واذا اسقط المربع الكائن من خط ا ح المشترك صار الباقي وهو المجتمع
من ضرب خط ا ب في خط ب ح اربع مرات مساو للباقي وهو المربع الكائن
من خط خ ح والمجتمع من ضرب خط ا ب في خط ب ح اربع مرات
مساو للمجتمع من ضرب خط ب ح في اربعة امثال خط ا ب ولكن اربعة امثال
خط ا ب خط ث فالمربع الكائن من خط خ ح ب ح والمجتمع من ضرب
خط ب ح في خط ث والمربع الكائن من خط ا د العام والمجتمع من ضرب
خط ا ب في خط ث وذلك ان سطح آ مربع وقد تمم خط ا د يتصل على نقطة
ب فالقطع العام الزاوية الذي يمر بنقطتي ب د يمر بنقطة خ ايضا ويصير

١١٧

ولنصف الخط الذي يوتر عليه الاعظم خط ت و وذلك ما اردنا ان نبين ٥

٥ ثم من بعد ذلك نبين ديه

ونو ملس ان الانوار المتساوية
الموضوعة على خط مستقيم
ان يمر متساوية واعظمها الذي
يوتر على العمود الذي يخرج من البصر
الى ذلك الخط ونضع ان نبين
ذلك فليوضع مربع ا ب ج د و
نوصل خط ب ج ولنرسم على مركز ب ج كه وبعد ب د قوس د ط ا
وهو من انا نمر بنقطة آ ونعمل قوس د ط مساوية لقوس ط آ فليقسم
خط ج د بخطوط متساوية كم كانت وهي خطوط ج ه ه ز ز ح ح د
ولنوصل خطوط ه ب ز ب ح ب فهو بين ان زوايا د ب ج ما قدار د
ح ح ز ز ه ه ج بين غير متساوية وخط ج د اطول من قوس د ط لان
خطي ج د م آ اطول من جميع قوس د ط آ فليس قوسها متساويا للشكل
الذي يحيط به قوس ط ل د وخطي ط ب ب د ولنعطف على قوس
ط ل د صفحة ولنجعل عليها نقطة ك ل ثم نرسم الصفيحة التي
سمته ج ولنعلم النقطة التي كانت علمت عليها نقطة مثل م ت ولنخرج
من نقطة سه خط سه له على خط سه ع على زوايا قائمة ونجعل خط سه
مساويا لخط سه ج ونصل خط ل ع ونخرجه الى جهة الافق الى نقطة
ق ونجعل خط له ق مساويا لخط ج د فهو يكون اطول من له سه و
ذلك انا قد قلنا فيما تقدم انه اطول من قوس ط ل د التي هي مثل خط
له سه ولنخرج على نقطة ق خط موازيا لخط سه ع وهو خط ق ذ ق
ولنوصل خطوط له شه له ت له ث ولنخرج الى نقطة خ ذ خطه
فنسبة زاوية ج ب د الى زاوية ح ب د كنسبة قوس ط ل د الى قوس
م د التي هي مثل نسبة خط سه ع الى خط سه شه التي هي مثل نسبة
خط له ق الى خط ق خ التي هي مثل نسبة خط ج د الى خط ح ج
ولان خط ح د الى زاوية ح ب د كخط ق خ من من زاوية ج ب د

١١٨

وكذلك السطوح

خط ز ح السطح مثل

خط ح ز وخط

ه ز مثل خط ذ ق

وخط ح ه مثل

خط ه ق فخطوط

د ح ح ز ه ح مثل ما مر من خطوط ط ف ح ح ح ز ذ ه ق وهو
مثل الخطوط ف ح ح ح ز ذ ه ق ه بعضها اطول من بعض لان نسبة د م
م ك ل ك ك ط ك اعنى بعضها اطول من بعض لان الزوايا التى عند المركز
بعضها اعظم من بعض ٥ وقد بين ارشميدس فى القول فى الكرة
والاسطوانة ان كل قطعة كرة فهى مساوية لمخروط الذى قاعدته قاعدة
القطعة وارتفاعه خط مساو للخط المحدود الذى الا يخرج من راس القطعة
التى قاعدتها مثل نسبة نصف قطر الكرة والعمود القطعة الاخرى من الكرة
الى العمود ارتفاعها جميعا الى عمود تلك القطعة الباقية فلتكن كرة
كرة اب ح وليقطع بسطح ما وهو سطح الدائرة التى يكون قطرها ح د
وليكن قطر دائرة اب ح وقطعه ه مركزها وليخرج خط ح ز
الى خط ز ب كنسبة خط ه ا الى مجموع الخطين ز ا وليكن هذا العمل ايضا
مخروطا نسبة خط ط د الى خط ز ا فقد بين ان قطعة ح ب د من الكرة
مساوية للمخروط الذى قاعدته الدائرة التى يكون قطرها ح د وسهمه
خط ط د فلا بد لذلك اذا اردنا ان نقطع الكرة المعلومة بسطح ما حتى يكون
لقطعتى الكرة احديها الى الاخرى نسبة مثل نسبة معلومة فعلينا ان نسبة المخروط
الذى قاعدته الدائرة التى على قطر ح د وارتفاعه خط ط د الى المخروط
الذى قاعدته تلك الدائرة بعينها وارتفاعه خط د ح معلومة كنسبة خط
ط د الى خط د ح وذلك لانه قد بين ان المخروطات التى تكون على قواعد
متساوية فان نسبة بعضها الى بعض كنسبة ارتفاع بعضها الى بعض فنسبة خط ط د
الى خط د ح معلومة والان نسبة خط ط د الى خط ز ا كنسبة خط ه ب
د مجموعا الى خط ب د يكون اذا فصلنا نسبة خط ط ا الى خط ا د كنسبة

119

خط سح الى خط رح واذ كان ايضا يلزم ان نسبة خط ح ت الى خط ت
واذ كان ايضا يلزم ان نسبة خط ح ت الى خط رح كنسبة خط ه آ الى خط ا ت
وخط ه آ هو مثل خط ب ه فقد بان صحة المسئلة على ما اصف وهو انه
اذا كان خط ا ت معلوم الوضع وكانت نقطتا آ ت معلومتي الوضع وكان
خط ه ت معلوم القدر فقد نقسم خط ا ت على نقطة ر ونصل له خطا آ
ت ح حتى يكون نسبة خط ط ر الى خط ر ح معلومة ويكون مع ذلك نسبة خط ط آ
الى خط ر آ كنسبة الخط المعلوم الذي هو ه ت الى خط ر ت ويكون مع ذلك
نسبة خط ح ت الى خط ر ت كنسبة ذلك الخط المعلوم بعينه الذي هو ه ت
الى خط ر آ وقد تبين ذلك
فيما بعد وذلك ان يمكن
2 كما بين ما قلناه بوجه
المولس هذا صار المسئلة الذي لم
يبينها في كتابه في الكسر اذا كان خط ا ب معلوم بالوضع وكانت نقطتا
آ ت معلومتي الوضع كيف نقسم خط ا ت على نقطة ح مثله اذا اخرجت
اليه خطا د آ ت صارت نسبة خط ح د الى خط ح ه كنسبة معلومة
وصارت ايضا نسبة خط د آ الى خط اح كنسبة خط ما معلوم الى خط
ح ت فصارت نسبة خط ه ت الى خط ت ح كنسبة ذلك الخط المعلوم
الى خط ح آ فلنفرض ذلك ونخرج خط د ت فانا نعمل على كل واحد من خطي
اح ت ط آ مساويا للخط المعلوم ونوصل خطا ح ح ح ط ونخرجا
الى نقطتي ر ك فلان نسبة خط د آ الى خط ا ح كنسبة الخط المعلوم
الذي هو خط ط ح الى خط ح ح وهي النسبة مثل نسبة
خط د آ الى خط اح يكون خط ط آ مساويا لخط ا ذ والذي ايضا
تبين ان خطي ر ت مساويا لخط ت ه وهو بين ان سطح ا ط الموازي
الاضلاع معلوم وذلك ان كل واحد من خطي ا ت اح معلوم والاثنية
خط د ت الى خط ح ت معلومة يكون نسبة خطي ك ا اح مجموعة
الى خط ت ح معلوم فنسبة المجموع من خطي ك ا اح

١٢٠

مجموع مربعي خطي زب جد مجموع مربع المربع الكائن من خطي زب بجد
مجموع مربع معلوم ولان نسبة خطي اج اد مجموع مربعي خطي جد زب
مجموع مربع نسبة خطي كا اد مجموع مربعي خطي طب بجد مجموع مربع زد ك
ان كل واحد من هاتين النسبتين من نسبة خط اد الى خط بجد
فيكون المجتمع من ضرب خطي دا اد مجموع مربعي خطي طب بجد مجموع مساويا
للمجتمع من ضرب خطي كا اد مجموع مربعي خطي زب بجد مجموع الكسر نسبة
المجتمع من ضرب خطي كا اد مجموع مربعي خطي زب بجد مجموع المربع
الكائن من خطي زب بجد مجموع معلوم فنجعل خط ال مساويا لخط دا
ونجعل خط سم مساويا لخط طب فنقطتا لم معلومتان فيكون
ان نقطة ك اذا كانت فيما بين نقطتي دا تقع نقطة م خارج نقطة.
ه لانا قد بينا ان المجتمع من ضرب زح في حه مساو للمجتمع من ضرب
سح في حم فيصير لنا المجتمع من ضرب لح في حم معلوما لان المربع
الكائن من خطي زب بجد مجموع معلوم ونخرج على نقطتي خط موازيا
لخطي ... وهو خط ... ونكمل على نقطة ... قطع خط ... على
زوايا قائمة وهو ... ونجعل خط ... مساويا لخط سح ونوصل
... النقطة ... معلوم الوضع لان نسبة خط
سح الى خط ... معلوم ... عمود ... على ... نقطتي لم ...
... خط ... ونسبة ... الى ... خط
... المربع الكائن من خط ... الى المجتمع
من ضرب ... خط ... المربع الكائن من
خط ... المربع الكائن من خط ... المجتمع من ضرب ...
... المجتمع من ضرب ... المربع الكائن
... المربع الكائن ... معلوم لان نسبة المربع الكائن
من خط ... المربع الكائن من خط ... معلوم نسبة المجتمع
من ضرب ... المجتمع من ضرب ... معلومة
ونسبة المجتمع من ضرب ... المربع الكائن من خطي زب بجد

١٢١

مجموع مربع الاول هو مثل المربع الكائن من خط سه ت معلومة وزاوية
سه ق ك معلومة وذلك انا نعمل زاوية قائمة ونخط قوس ق ر ه معلوما ان =
نقطة سه ح على خط محيط نقطع ناقص معلوم الوضع ونسطح ا ط ما ونسطح
سه ك على ر ح لكن سطح ر ك فالمجموع من سطح ط ح ك ط = مساو لمجموع
من سطح سه ت ل سه ح ق وخطا ك ح ح ط معلوما الوضع ونقطة ت
معلومة فان عملنا قطعا زائدا يكون خطا ك ح ح ط الخطين اللذين لا يلقيانه
ومن نقطة ت معلومة فانه من نقطة سه ه نقطة سه ح على خط محيط
نقطع زائد معلوم الوضع وهي ايضا على محيط قطع ناقص معلوم الوضع
فنقطة سه ح ر معلومة وعلى
خط ا ب نقطة ح ر
معلومة ونسبة خط ه
= الى خط ب ح
لنسبة الخط المعلوم الذي هو مثل
خط ا ح الى خط ا ح معلومة
ونسبة خط ه ت الى خط ب ح معلومة فنقطة ه
ايضا معلومة وذلك نعلم نقطة نقطة د ونزيد ذلك بين
٥ فنريد ان نبين كيف نخط خطا يكون مثلا وهو الخط معلوم فلكن
الخط المعلوم خط د ه ونريد ان نخط خطا يكون من حره ونمس مثل
الخط المعلوم او مثل هو وهو الموضع فليكن خط د ا قائما
على خط د ه على زوايا قائمة ولنفصل خط ا د على استقامة ونصل
ولنعلم على خط د ت نقطة ما ونصل ونخط ت ولكن
خط د ا بنسبة امثال خط د ت ونخرج اه ح ونخرج خط ح ت موازيا
لخط ا ه والان خط ا د بنسبة امثال خط د ت يكون خط ب ا مثل ه
ولما كان خط ا د لخط ب ح انما مثل خط د ه فيكون عدد حدها
الخط الاول اردنا وهو خط ب ح فان اردنا ان نخط خطا اخر ادا كان
خط ت ح معلوما حتى يكون خط ب ح مرة ومن مثل ذلك الخط فانا
نخرج خط ب آ على زوايا قائمة على وضعها ونعلم عليه نقطة ك اين

١٢٣

ما وقعت ونجعل اذ مثله امثال د ت ونخرج خط اح ونفصل
د ه موازيا لخط ب ح فيصير خط ب ح مرة ونفصل خط د ه فان
اردنا ان يكون مرة ونصفا صيرنا خط اذ نصفه اصغر من خط د ت فذلك
تلك الاشياء التامة التي عملنا بابعادنا فليس الخط المعلوم خط ب ح ونريد
ان نضعفها لخط د ه حتى يكون خط ب ح مثلا واجزاء لخط د ه كانا ه
اردنا سبعة اشياء بثلثة اخماسه حتى يكون نسبته اليه نسبة سبعة الى خمسة
فنخرج خط ا ب على زاوية قائمة ونفصل خطوطا متساوية بعضها لبعض
ونجعل خطوطات سبعة منها ونفصل خط اذ خمسة منها ونصل خط اح
ونخرج خط د ه موازيا لخط ب ح فيصير خط د ه هو الخط المطلوب
ولذلك ايضا ان اردنا ان نجد خطا آخر له الى الخط المعلوم نسبة اخرى معلومة
اى نسبة كانت فانا نعمل هذه الاشياء بابعادها
ثم نبين من بعد ذلك ذو مرلمس كيف نجد مكعبا
يكون ضعف مكعب ببرهان هندسي ه
فليكن الخط المعلوم خط آ ونريد ان نخرج
خطا آخر يكون مكعبه ضعف مكعب خط آ مثلا فنضع خطين متساويين على زاويا
قائمة وهما خطا ح د و ه ر ونفصل كل واحد من خطي ح ح ح د ر ح
خط آ ونجعل خط ت نصف خط آ ونفصل كل واحد من خطي ه ح
مثل ربع خط ت ونخرج خطا ط د ه ط على زوايا قائمة ونجعل خط
د ك مساويا لخط ح ك وهو مثل ان خط د ه يكون ايضا مساويا لخط ه ك
آ مثلا خط ت فنعمل على خطي ح م ه ق نقطا ما كثيرة متساوية
وهي نقط ل م ع ق ونخرج من نقطتي ل م من خط ح م عمود الى نقطة
م ل مثله ونخرج من نقطة ع ق اعمدة على نقطة خط ح ق وهي خطوط ط
ع شه ل ق ه ر ه والخط الذي يخرج من نقطة د الى نقطة ج مساويا
لخط ح ك فنعمل نقطة د مركزا ونرسم ببعد مثل ح ر د خطا محيطا
وارح ونعلم على موضع قطعها خط ل ق نقطة ق ولذلك ايضا نرسم
على مركز د ونبعد ح م خطا محيطا وارح ونعلم على موضع قطعها لخط م
س ه نقطة س ق والذي نرسم بالمسطرة التي نعطف خطا بر نقطة ك ذ س ق

١٢٣

وسام السطح الذي يسامت على يمين لكه وهو خط كـ ز شبه الدوائر ايضا
نعمل بالخط الاول منه نرسم على م كرة ه ونصد ع خطا محيطا مار ونقطع
خطوط شبه على نقطة شبه الدوائر بعد المركز ه بعضها والبعد لكل واحد
من خط ونقطة ز ونرسم خطين محيطين متوازيين والسطح فنقطة
نه ز ه على نقطتي نه ك والدائرة نرسم بالمسطرة التي عليها خط ما من نقطة ط شبه
نه ز ه وهو خط ط شبه نه ز ه خطا كـ نه شبه ط شبه نه ز نقطع احدهما
الاخر على نقطة ما فلتقع على النقطة نه ونخرج عمود نه ك على خط ط م
فاقول ان ضلع خط آ ضعف ضلع خط نه ك وذلك ان خط هـ ك
مساو لخط الذي يخرج من نقطة د الى نقطة نه فالذي يكون من لـ ج سا ح د
اربع مرات مع المربع الكائن من د ك مساو لمجموع الكائن من خطي د ل لـ نه
ولسقط المربع الكائن من خط د ك المشترك فالباقي الذي هو المجموع من لـ ج
ح د اربع مرات الذي هو مثل المجموع من ضرب آ ما ح ك مساو لمربع
ل نه فسبة خط آ الى خط ل نه كسبة خط نه ك الى خط لـ ج والدائرة
سهل بين المجموع من ضرب خط ت في خط ح ق مساو لمربع خط ق نه الذي
هو مثل المجتمع من ضرب ت في خط ل نه مضروب خط ت في خط ل نه مساو
لمربع الكائن من خط ف نه الذي هو مثل المربع الكائن من خط ح ك فنسبة
خط نه ك الى خط لـ ج كنسبة خط لـ ج الى خط ت وقد بان ان نسبة
خط نه ك الى خط لـ ج كنسبة خط آ الى خط نه ك بين خطي آ ت خطا
لـ نه لـ ج متواليان على نسبة
خط آ الى خط نه ك وخط آ مثل
ثلاث مرات مكعب خط آ مثلا
مكعب خط نه ك فالخط المطلوب
هو خط نه ك ولذلك ايضا
نعمل به الباقية
المعلومات بما كان وبين البراهين لنه خطي ط شبه نه ز فهو كـ نه شبه
مما قطعها يحدد عام الزوايا ويمكن وليصدهما من نقطة نه فالقطعان
نقطعان بعضهما الاخرى لا محالة ن وهو يعد الكائن المتوسطان على خ ح   ما

١٢٤

وهي لو كانتا دائرة ا ب ج د والكس منها فظاهر يقطع احدهما الاخر ب
على زوايا قائمة ومحاذات ج د المفصل من الرابع موسس متساوين
ومحا موسا ك د د ك والمفصل خط ب ك فاقول ان خطي
رك طت متماس خطي اك طح على نسبة والامر قال نسبة خط اك
الى خط طر كنسبة خط ما الى خط طت بين فاقول ليس نسبة
خط رك ايضا الى خط طت كنسبة خط طت الى خط طح فلنخرج
من نقطة ك الى الخطات عمود ك ل ضمنه خط سط الى خط طح م
كنسبة خط سل الى خط لك مما خط سل بموصل خط اك واما خط
لك بموصل خط رك لان قوس رد مساوية لقوس دك وخطار كم
ك ل عمودان فنسبة سا الى خط طح كنسبة اك الى طر التي قلت
انها نسبة رك الى طت بخطوط اك طر ط
طح الاربعة متوالية على نسبة ه والعكس
ايضا واسر علينا اسحدد الملس منها فظاهر
احدهما قاطع للاخر على زوايا قائمة ومحاذيا
اسحدد والمفصل من الرابع نسبا متوالية
متساوية بعضها لبعض ... ان ب ج ح ط ونخرج الى خطات العمود
ب ح ل ط م والمفصل عند نقطة د من الرابع الاخر من الذي بخ
متساوية لقوس د ك ب ح ط ح يكون على ما مساويا للعدد
ايضا وهي شي د ك ..ح والمقطع الخط الذي يصل بين نقطة
ت ونقطة ف خط رك على نقطة ت والمقطع الخط الذي يصل بين
نقطة ت ونقطة س خط ح ل على النقطة ق والمقطع ايضا الخط
الذي يصل بين نقطة ت ونقطة ع خط طم على النقطة ر فعند ذلك
مما قلنا فيما تقدم ان خطي رك كت متوالان فيما بين خطي اك
ك ق على النسبة ولكن ايضا يكون خطا ح ل ل ت متواليا بين
خطي اك ل ح على النسبة ويكون خطا طم م ت متواليان فيما بين
خطي ام م ر على النسبة فان جعلنا الا عمدة التي تخرج من الشكل
وتعمل عليها نقطا كما عملنا نقط ف ق ر وخططنا السطوح

١٢٥

اين ينعطف من جهة السطح كلها خطا
وهو خط س د ب لهو بين انه ان يعلم
عليه نقطة ن واخرج منها عمود ن ك
على خط ا ت صار خطا ر ك د ت مما
خط ا ك ك ع مساويا على نسبة ه فاذا سبب هذين الاشباه
فانا نجعل خطا ا معلوما ونجعل نسبة خط ت الى خط ح كمعلومة ونريد
ان نجد خطا مثل خط سه حتى يكون نسبة المكعب الكائن من خط ا
الى المكعب الكائن من خط سه مثل نسبة خط ت الى خط ح فنخط دائرة ح ز ك ر
ونجعل نصف قطرها مساويا لخط ا ونكون منها القطعة التي قاطع
نصفها الآخر على زوايا قائمة وهما خطا د ر ط ح ولنرسم خطا على ما
وصفنا وهو ر ك ط ونعمل نسبة خط د ه الى خط ه ل كنسبة خط ت
الى خط ح ونصل خط د ل ونخرجه حتى يلتقي من الخط الاول وصفنا الى
نقطة ك ونخرج الى خط د ر عمود ك ع ونخرجه الى نقطة ل ونصل
خط د ك ونكون خط سه مساويا لخط ه ع فاقول ان الخط
المطلوب هو خط سه لانه قد تبين من الاشكال التي قدمناها ان خطي
ل م م ن متوالية على نسبة خطي د م م ح على نسبة ولما كانت اربعة
خطوط متوالية على نسبة فان نسبة الاول منها الى الرابع كنسبة المكعب الكائن
من الاول الى المكعب الكائن من الثاني فنسبة خط د م الى خط ح م
التي هي مثل نسبة خط ه د الى خط ه ر التي هي كنسبة خط ت الى خط
ح هي كنسبة المكعب الى المكعب الكائن من خط د م الى المكعب الكائن
من خط ل م التي هي نسبة المكعب الكائن من خط د ه الى المكعب الكائن
من خط ه ع وذلك كنسبة المكعب الكائن من خط ا الى المكعب الكائن
من خط سه فان اردنا ان نجد خطا اخر حتى يكون نسبة المكعب الذي
يكون منه الى المكعب الاول يكون من خط ا مثل نسبة المكعب الاول يكون
من خط ت الى المكعب الاول يكون من خط ح لهو بين انا ان جعلنا المربع
الكائن من خط ه ا مساويا لمجموع مربعي خط سه الخط الآخر صار ذلك
الخط المطلوب ذلك الخط الآخر وذلك انه يصير نسبة ذلك الخط الى خط ا

١٢٦

كنسبة خط آ ب للضلع القائم
من خط آ ب الى الضلع ح ر ك واذا يد
كان ذلك كذلك فانا نعمل نسبة
مستقيمة اليها على زاوية
لسطح يكون مثل ذلك الى كل
واحد من الخطوط المعمولة بعدد
خط ب ك انما على خط ا ت على زاوية قائمة ونعمل خط ب ل قائما
خط ا ت فاذا اخذنا من نصف دائرة خط ت س ا د [illegible] الشكل
صار لنا من الصورة مثلث قائم الزاوية واحد اضلاعه زاوية
ت قائمة وخط ا ت مساو لخط ب ك وعند خروج [illegible] من نقطة التي هي
راسه الى ما تحت [illegible] خطا ب ز [illegible] على نقطتين مشتركتين
بها وخط [illegible] منها قطعة من خط محيط [illegible] وخط ب ر د الخط
الذي تقدم ذكره واما [illegible] ذلك فانا
نصل [illegible] نعمل ايضا دائرة ه
ا ب ح د التي فيها نظائر ا ت ح د
اللذان [illegible] على زاويتها
قائمة ثم نخط ايضا خط ا د [illegible] ر د على [illegible] الصورة المتقدمة ونخرج
خط ا د [illegible] ونعمل مثلث ا ب ح د الذي خط ا ت منه مساو لخط ب ح د
وزاوية ت منه قائمة ونخط ايضا [illegible] العظمى
خط ب [illegible] منها قطعة من المحيط ما الواسع وخط ب ل [illegible] الخط الذي
تقدم ذكره ولكن خط د معلوما ولتكن نسبة ا ل ه الى الخط ر معلوم ونفصل
خط ا ح مساويا لخط د و [illegible] على نقطة ح خطا موازيا لخط ب ح د
[illegible] خط ط و [illegible] نسبة خط ا ل الى الخط [illegible] كنسبة خط ه الى الخط ر
[illegible] ونصل خط ا ح [illegible] على خط ل خطا موازيا
لخط ب ح [illegible] وهو خط م ل ر ونصل من نقطة ر ونقطة ا خط ا ر فقد
تبين من الاشياء التي قدمت ان نسبة خط ا م الى خط م ل التي هي نسبة
خط ا ح الى خط ح ك التي هي مثل نسبة خط ه الى خط ر كنسبة المكعب

١٢٧

الكائن من خط ا م الى المكعب الكائن من خط م ر التي هي مثل نسبة المكعب الكائن
من خط ر ح الى المكعب الكائن من خط ح ع وخط ا ن مساو لخط د فاذن الخط المطلوب
هو خط ح ع فان كان خط د اطول من خطات نسيا فيقسم قسمين
بعده بقسمين ولا نزال نفعل كذلك حتى يبقى لنا خط ما اقصر من خطات تسهل
الوزن فيبقى لنا خط اح د يعظم اضعاف الخط المعلوم كخط ل ا وما شد
حطاما يكون اضعافه كخط ع ح فمثل ذلك الاضعاف فيكون ذلك الخط هو
الخط المطلوب وان كانت غلطنا ح ع في المقالة الاولى من قوس
ـه فانه لم يكن من الامرين فرق وذلك لان العمل عمل واحد بعينه والتركيب
البرهان لانه يكون نسبة خط ه الى خط ي كنسبة المكعب الكائن من
خط ا م الى المكعب الكائن من خط م ع و
لنذكر اعمالنا الشكل التي نعمل على الخط الخطير
الى التي تعمل على الاجر والاسطوانة الى الاسطوانة او شيء مما يعمل من الشكل او
الاسطوانة او ما شبه صورتها فان العمل والبرهان شائع في ذلك
واحد ه قد بينا ان قول الاستاذ انه يجد خطين بين خطين
معلومين على توالي الاربعة الخطوط على النسبة وذلك انا نجعل الخطين المعلومين
اللذين نريد ان نجد بينهما خطين على توالي الاربعة على النسبة واحدا فنجعل احدهما كخط
ونضع خطا ان خط لكان وهو خط ح ز والاخر خط ا ل ت كنسبة
المكعب الكائن من ح ز الى المكعب الكائن من د كما بينا ونضع خط
نجعل خط ح د خط ثالث مناسب لهما وهو خط ه ونجعل د ه خط ثالث
مناسب لهما وهو خط ى فخطوط ح د ه ى متوالية على نسبة
فنسبة خط ح الى خط ى كنسبة المكعب الكائن من خط ح الى المكعب
الكائن من خط د التي هي نسبة خط ا الى خط ت فنصير نسبة خط ا
الى خط ع كنسبة خط ح الى خط ه وخطوط ح د ه متوالية على نسبة
ونسبة خط ا الى خط ت مثل نسبة خط ح الى خط ى ونسبة خط ا الى خط
ط مثل نسبة خط ح الى خط ه فيكون نسبة خط ط الى خط ت كنسبة خط
ه الى خط د التي هي مثل نسبة خط د الى خط ه التي هي مثل نسبة
خط ع الى خط ط فخطوط ا ع ط ت متوالية على نسبة فقد وجدنا

مما بين خطي آ ت خطوط ساعات الشمس فعلها عمل الشمالية وبها خلا في ط هذا
اجزاها وحده من كتاب دستور طلسن فائم اما الجزء تمت بعون الله تعالى
والحمد لله وحده وصلى الله على سيدنا محمد
وآله الطيبين الطاهرين وسلم
تسليما كثيرا

بسم الله الرحمن الرحيم رب يسر وأعن

قال أحمد بن محمد بن كثير الفرغاني إن المتقدمين من العلماء
بحساب النجوم لما أدركوا علم حركات الفلك وما يوصف به بالبراهين
على البسط والقياس وإن كان أكثر ما استدلوا به من الأسباب المقاييس الآلة
التي تسمى ذات الحلق إذ كانوا عملوها على ما يشبه هيئة الفلك في استدارة
حركاته وخاصة الحركتين الأوليين أعني بالحركتين الأوليين حركة
الفلك الأعظم المدبر للأفلاك كلها من المشرق إلى المغرب وحركة الكواكب
كلها في الفلك المائل من المغرب إلى المشرق وهذا الفلك هو فلك البروج
وقطباه يدوران على قطبي الفلك الأعظم والعلة في صحة ما استدلوا به
من هذه الآلة والصحة لا تكمل المعرفة بالهندسة وأما هم عملوا الآلة التي
تسمى الاصطرلاب مسطحة على غير خلقة الفلك استدلوا بها على كثير من حركات
الفلك في اختلاف الليل والنهار في أقاليم الأرض وأجزاء الأفق
في صنعتها والقياس بها على ما يلزم معرفته حاجة إلى معرفة العلة في هيئتها وصحة
دلالتها ولم يبلغنا أن أحدا منهم بيّن ذلك ولا رسمه في كتاب وكان
استعمال الناس إياها على جهة التقليد وأن ما يخرج بها من حركات الفلك
موافق لما يخرج بآلة ذات الحلق وما يخرج أيضا بالحساب وإن لم يكن في
ذكر برهان على صحتها ولا دليل على علتها وكان مما ذكرنا من اشتباه
الأمر فيها لما ادعاه عدة ممن يوصف بالعلم في زماننا إلى أن زعموا أنها ليست
معمولة على الحقيقة وتوهموا أن يكون المبتدئ لصنعتها اعتسف عملها
على غير صحة وبعض من يصف اقتدر بلوغ المعرفة في ذلك سلاك الطرق بأن
على صنعة الاسطرلاب على ما تصح علته عندهم وعلى خلاف ما لم يزل

2 D̲iyūqlīs Here and in some other places (4 [bis], 5, 20^{1}, 186, but not 6, 20^{2}, 125, 136, 149, 252) the ms. displays d̲ for δ in proper names and other words transcribed from the Greek. I have restored d̲ in every case, since it represents Greek pronunciation contemporary with the translation (on this pronunciation see Schwyzer I p. 208). The best treatment of this and similar phenomena in Arabic transliteration is Daiber, *Placita Philosophorum* pp. 52 ff. ("Zu den Konsonanten in nachklassischen Griechisch").

On the book's title see Introduction p. 3.

3 Pythion (Fūt̲yūn) Nothing is known about this man from any other source, but the name is certain. Πυθίων is a theophoric name of a common type (cf. Heron, Hephaestion, Dion), and the name is not rare, but occurs in many parts of the Greek world. Without attempting an exhaustive survey, I have noted examples from Athens, Megara, Elis, Boeotia, Thessalonike, Amphipolis, Parion, Samothrace, Mytilene, Cos, Rhodes, Magnesia on the Maeander, Iasos, Miletus, Colophon, Patara and Side. But it is significant that the name was very common indeed on Thasos: in the prosopography of Thasos, *Etudes Thasiennes* V pp. 253-311, it is one of the two most common names (44 examples, none of which can with good reason be identified with "Pythion the geometer"). The name was so popular on Thasos because Pythian Apollo was the principal deity of the community (on the connection of Greek names with local deities see Letronne, "Etude des noms propres grecs" pp. 35-49; cf. L. Robert, *Opera Minora* I p. 460: "ce serait une étude attachante... de rechercher dans quelle mesure on peut conclure d'un nom théophore à l'existence d'un sanctuaire du dieu dans le pays"). The temple of Apollo, the Pythion, was the chief shrine on the acropolis of Thasos. (The other most common name on Thasos, also with 44 examples in the prosopography, was Herakleides: Herakles was, besides Pythian Apollo and Athena Poliouchos, the most important deity in ancient Thasos).

allad̲ī min ahl T̲āsīs This is a translation of ὁ Θάσιος (the misspelling T̲āsīs for T̲āsus is probably due to the translator's ignorance of the noun from which the adjective is derived). Cf. Apollonius *Conics* IV Introduction (ms. Marsh

667, 70^r marg.): Ūdīmūs allaḏī min ahl Barġāmūs = Εὔδημον τὸν Περγαμηνόν; Galen, *On Cohesive Causes* (ed. Lyons p. 54): Āṯīnāūs allaḏī min ahl Āṭālyā = 'Αθηναῖος ὁ 'Ατταλεύς (for the latter see Galen, ed. Kühn XIX p. 347).

Conon (Qūnūn) Conon of Samos, the well-known mathematician and astronomer, dated by his naming of the constellation *Coma Berenices* in honor of the consort of Ptolemy III Euergetes, probably in 246 B.C. (hence Pythion too can be dated to the mid-third century). Archimedes mentions Conon more than once as a former friend and mathematical correspondent who is now dead (*Sphere and Cylinder* I Pref., II Pref., Heiberg I pp. 4 and 168; *Spirals* Pref., Heiberg II p. 2; *Quadrature of Parabola* Pref., Heiberg II p. 262). On Conon in general see Rehm, RE; Bulmer-Thomas, "Conon".

The same transliteration of the name is found in ms. Marsh 667, 70^r (Apollonius, *Conics* IV Introduction).

4 Zenodorus The readings of the manuscript, ʾbyūḏām-s (here) and ʾynūḏām-s (5), do not correspond to any possible Greek name. The emendation to Zīnūḏurus (or possibly Zīnūḏūrus) is certain: for confusion between alif and zāy or rā in this script cf. 46, where the ms. (p. 115 line 4) has "AT" for "RT". Zenodorus was a mathematician of the early second century B.C., best known for his work *On Figures of Equal Boundary*, parts of which have been preserved by Theon and Pappus. He is also, however, known as an astronomer (listed as one of οἱ περὶ τοῦ πόλου συντάξαντες in ms. Vat. gr. 381, published by Maass, *Aratea* p. 123). Both interests are confirmed by this passage, where Zenodorus is referred to as "the astronomer", but he asks a question of mathematical import (though he may also have been interested in sundial theory, cf. 21). He was associated with the Epicurean philosopher Philonides, who was also an acquaintance of Apollonius: see my detailed discussion of the evidence for the dating and identification of Zenodorus ("The Mathematician Zenodorus"), where I also tentatively suggested that he was a member of a prominent Athenian family from the deme Lamptrai, among whom the name Zenodorus was hereditary. On the importance of this passage for the dating of Diocles see Introduction p. 2.

Arcadia (Arqāḏiyā) However one translates in detail (see next note), one must conclude that Diocles was living in Arcadia and that Zenodorus came to visit him there. See Introduction p. 2.

was introduced to us I have emended "fī-hā" of the ms. to "la-nā". If the ms. reading were kept, we should have to translate "was appointed to a teaching position there", which is less appropriate, though perhaps possible.

5 we shall make use... predecessors This statement is amply borne out by Prop. 1, where a number of theorems in conics are silently assumed. See 40, 41, with notes.

6 ʿamila-hā My translation ("solved practically") is meant to imply, not necessarily that Dositheus actually constructed a parabolic mirror, but that he stated the focal property of the parabola, perhaps without giving a formal proof. See Introduction p. 16.

Dositheus Our best information on this man comes from the preface to Archimedes' *Quadrature of the Parabola* (Heiberg II p. 262), where Archimedes, addressing Dositheus, announces that, hearing that Conon is dead and that Dositheus was a friend of Conon and versed in geometry, he has decided to send the following treatise to him instead of to Conon. Dositheus is also the addressee of Archimedes' treatises *On the Sphere and Cylinder*, *On Conoids and Spheroids* and *On Spirals*. He is given as the authority for the dates of rising and setting of various stars in several ancient calendars, and is named as the author of other lost works dealing with astronomy and the calendar (see Hultsch, RE). The information here, that he recognized the focal property of the parabolic mirror, is new, but it is not surprising that he investigated conic sections, in view of the subject matter of the treatises addressed to him by Archimedes, and the interest of his friend Conon in the field (see Apollonius, *Conics* IV Pref., Heiberg II p. 2).

7 We have set out The reference is to Prop. 1 (38-67)

8-9 Here is enunciated the focal property of the paraboloid of revolution. It is proved in the first part of Prop. 1 (38-50).

8 qaṭʿi ʾl-makrūṭi ʾl-qāʾimi ʾl-zāwiya This is a literal translation of τομὴ κώνου ὀρθογωνίου ("section of a right-angled cone"), the pre-Apollonian term for parabola (see e. g. Archimedes, *Conoids and Spheroids* Pref., Heiberg I p. 246, 16, cf. Eutocius, *Comm. on Apollonius* I, Heiberg II pp. 68-70). It is the only term for parabola used by Diocles. On the significance of this see Introduction p. 9.

al-ḵaṭṭi ʾllaḏī yaqsimuhu bi-niṣfayni This, which is the standard term in this text for the axis of the parabola (cf. 38), probably translates ἡ διχοτομοῦσα. Compare Apollonius, *Conics* II 10 (Heiberg I p. 208, 1), τῇ διχοτομούσῃ διαμέτρῳ (which refers, however, to a hyperbola). It is noteworthy that Diocles uses neither the old Archimedean term διάμετρος nor the new Apollonian term ἄξων (cf. note on 9, pp. 141-2).

9 whose distance from the surface The distance must of course course be measured along the axis.

al-ḵaṭṭ allaḏī taqwā ʿalayhi al-aʿmida etc. This translates [τῆς εὐθείας] παρ' ἣν δύνανται αἱ πρὸς ὀρθὰς ἀγόμεναι vel sim. Cf. Archimedes, *Conoids and Spheroids* III (Heiberg I p. 272, 16-17), παρ' ἃν δύνανται αἱ ἀπὸ τᾶς τομᾶς; Apollonius, *Conics* I 15 (Heiberg I p. 60, 5) παρ' ἣν δύνανται αἱ ἐπὶ τὴν AB καταγόμεναι τεταγμένως. This use of "qawiya", as noted by Nix, *Apollonius* p. 15, is a calque on the Greek δύνασθαι, which can only be paraphrased in English (e. g. "be equal in square to"). A δύναται BΓ means "A^2 = BG". It is combined with παρά only in this expression παρ' ἣν δύνανται κτλ, which is the standard way of expressing "the parameter" for all three sections in Apollonian conics, so that it is even used as a noun by Pappus *(Collection* IV 65, Hultsch I pp. 278-80), ὑπερβολήν, ἧς παρ' ἣν δύνανται ἔσται ἡ λοιπὴ εὐθεῖα, "a hyperbola, whose parameter will be the other straight line". However, the expression is pre-Apollonian, at least for the parabola, as was already known from a single occurrence in Archimedes (quoted above), and is confirmed by the present passage. On the expressions for the parameter of the parabola and the explanations for them see Introduction pp. 6, 7, 13.

al-aʿmidat allatī tuḵraju ilā ʾl-sahmi This translates αἱ κάθετοι οἱ ἀγόμεναι ἐπὶ τὸν ἄξονα, cf. Archimedes, *Conoids and Spheroids* XII (Heiberg I p.312, 6-7), τᾶν...καθέτων...τᾶν ἀγομενᾶν ἀπὸ τᾶς τομᾶς ἐπὶ τὰν ΑΓ. "sahm" is the standard Arabic translation of ἄξων, "axis", e.g. Apollonius, *Conics* I Definitions (ms. Marsh 667, 6^{v} lines 20-21), "wa-usammī al-k̲aṭṭ al-mustaqīm... sahman" = ἄξονα δὲ καλῶ... εὐθεῖαν, ἥτις κτλ. Diocles' use of ἄξων here at first sight seems to conform to the practice of Apollonius, who uses it for the axis of the sections as well as of the cone, and not to that of Archimedes, who always uses διάμετρος for the axis of the sections. However, Archimedes does use ἄξων for the axis of a conoid (an example of both usages is *Conoids and Spheroids* XII, Heiberg I p. 308, 27-28, ἄξων δὲ ἔστω τοῦ κωνοειδέος καὶ διάμετρος τᾶς τομᾶς ἁ ΒΔ), which is exactly parallel to the situation here, for Diocles is referring, not to the parabola, but to the paraboloid of revolution.

10 Reading and sense are uncertain. One would expect rather: "the greater the increase in the conic section, the greater the increase in the surface". But the Arabic makes a kind of sense, if one removes the words "ʿalā qiṭʿati dāʾiratin", ("to a segment of a circle"), which are meaningless in context. It is possible that Diocles wrote something like: "if the above-mentioned conic section is increased (in area) by a given amount, the surface will be increased in the same proportion, since the first is to the second as the radius of a circle to its circumference", and that this has been garbled in translation or transmission.

yuzādu The imperfect is odd: one expects the perfect in both protasis and apodosis, as required by the grammarians (see Wright II p. 14), and as found in the similar expression 52 "kullamā kāna al-basīṭ aʿẓam kānat al-šuʿāʿāt... akṯar".

12 from a spherical surface the rays are reflected to a straight line See Prop. 2 (68-77), where Diocles shows that all rays parallel to a given diameter of a spherical mirror are reflected reflected through one quarter of that diameter (AH in Fig. 2)

people used to guess that they are reflected to the center Exactly the same information is provided by the author of the Bobbio Mathematical Fragment (*Mathematici Graeci Minores* p. 88), who says that the ancients supposed that burning takes place about the center of a spherical mirror, but Apollonius in his work on the burning-mirror proved that this was false and showed where the burning would take place. It is likely that he is in fact referring to this work of Diocles (see Introduction p. 20). These "ancients" cannot be more precisely identified. The only other relevant text is the second part of Prop. 30 of the *Catoptrics* ascribed to Euclid (Heiberg pp. 340-42), which is an absurd "proof" that burning will occur at the center of a spherical mirror. On the date and nature of this text see Lejeune, *Recherches* pp. 112-36.

13-14 See Prop. 3 (78-96), in which Diocles shows that all rays parallel to a diameter of a spherical mirror reflected from a 60° segment of the sphere pass through a small section (less than 1/24th) of that diameter.

13 <u>place (mawḍiʿ)</u> This expression, rather than "point" (nuqṭa), is used because rays reflected from a spherical mirror of arc 60° pass through a small section of a straight line (see preceding note). The Bobbio Mathematical Fragment uses the phrase περὶ τίνα τόπον in the same context (*Mathematici Graeci Minores* p. 88, 11).

15 <u>later</u> 53-67, q. v.

16-37 This section is full of textual and historical difficulties to which I have no solution, or only a tentative one. I have little confidence in the text and translation presented for 22-27, and am doubtful about several other places. I suspect that the source of some of these difficulties is the failure of the Arabic translator to understand the Greek original. But for all its uncertainties the section is of great historical interest. If I interpret him correctly, Diocles says that the solution to Pythion's problem enables one to construct a sundial "without a gnomon", which indicates the hour by "burning a trace". I can explain this only as an application of the surface described at 53-59, which is produced by the revolution of a parabola such that its focus describes a semi-circle in the plane which is perpendicular to the plane of the parabola and contains the axis of the parabola. Then, if the surface thus generated is constructed out of a reflecting material, and set up so that its axis of symmetry points towards the sun at culmination, at equinox (when the sun is in the celestial equator), at any instant of the day the plane passing through the sun and the center of revolution of the surface (point E in Fig. 1) perpendicular to the focal semi-circle will always intersect the surface in a parabola. Thus the rays in that plane will all be reflected to a point on the focal semi-circle. If, therefore, a narrow strip of burnable material, such as wood or papyrus, were somehow positioned along the focal semi-circle, a line would (theoretically) be gradually burned along it by the reflected rays as the sun progressed during the day, and one could tell the hour by the progress of the burning (the strip could be divided into twelve equal parts and marked with the hour numbers).

The above interpretation fits the description at 16-17, including the statement that the mirror does not need to be turned to face the sun, and I am fairly certain that it is what Diocles intended. However, apart from the practical difficulties of construction, there are theoretical objections which make this type of sundial an impossibility. The most obvious of these is that the daily path of the sun lies on a great circle only twice a year, at the equinoxes. At other times it lies on a small circle, and however the mirror were positioned, the rays which intersect it in a parabola at any given moment would not be parallel to the axis of symmetry at any time during the day, and thus would not be reflected to the focal semi-circle. A further objection is that it is doubtful whether the rays in a single plane (as defined above) would be enough to cause burning when reflected to a point, unless the mirror were of incredible size.

Nevertheless, Diocles seems to discuss this "sundial" as a serious alternative to the conventional sundial with a gnòmon, saying that it avoids the objection which has been brought against the latter, namely that it entails the assumption that "every point on the earth can be treated as the center of the earth", i. e. that the earth's radius is of negligible size compared with the sun's distance, or that solar parallax can be ignored. Both modern and ancient sundial theory does indeed ignore solar parallax, and rightly, since it amounts to only a few seconds of arc at maximum. (See, e. g. Ptolemy, *Almagest* I 6, Heiberg I p. 20: one of the proofs that the earth has the ratio of a point to the heavenly bodies is that gnomons in any place on earth behave as if they were at the earth's center). However, in the time of Diocles the sun's distance had not been accurately determined (see note on 20, p.146), and it was still possible to argue (19) that the distances conventionally assigned to it might be too large, and that solar parallax is not negligible, hence sundials with gnomons are inaccurate. Diocles himself admits that solar parallax is negligible (22-24), but agrees with those who say that it is better to avoid this assumption (21), and maintains that his mirror sundial, which does not use the assumption, is theoretically more accurate than the gnomon type (24).

16 gnomon For the various terms for "gnomon" in Arabic, including "miqyās", the only term used in this treatise, see Schoy, *Gnomonik* p. 5.

17 above-mentioned figure The figure (šakl, cf. 8) is that mentioned in 15, and described, if I am right, at 53-59.

18 the astronomers Though he is thinking primarily of sundial theory, it is probable that Diocles is referring to all branches of astronomy. For in his time theoretical astronomy was mostly "spherics" of the kind represented by the extant works of Autolycus and Euclid, in which the earth is indeed treated as a point (e. g. Euclid, *Phaenomena* Prop. 1, especially p. 12 lines 9-10 (Menge). Even in the developed astronomy of the *Almagest* the effect of parallax is neglected in all topics except the position of the moon and eclipse theory, and Ptolemy devotes a whole chapter (I 6) to the thesis that the earth has the ratio of a point in relation to the heavenly bodies. Except for the very special case of Archimedes in the *Sandreckoner* (on which see Neugebauer, *History of Ancient Mathematical Astronomy* II pp. 644-45), we know of no case in which a Greek astronomer took parallax into account before Hipparchus, who was the first to determine the distance of the moon in terms of the earth's radius with reasonable accuracy (see Toomer, "Hipparchus on the Distances of the Sun and Moon"). Proclus (Hypotyposis IV 54, ed. Manitius p. 112) says explicitly that solar parallax is neglected by writers on sundial theory (τοῖς γνωμονικοῖς) and on analemmata.

20 30 million stades... 50 million stades The manuscript has "thousand" (alfa) instead of "million" (alfa alfi) in both cases. I have corrected the text as if it were a scribal error (haplography), but since the error occurs twice, it is perhaps due to the translator. The Greek must have been ,γ' (,ε') μυριάδες σταδίων, "3000 (5000) myriads of stades"; this can be read as "3 (5) myriads of stades" by simple omission of a diacritical mark, and since the translator probably had no conception of the size of a stade, that would be a plausible reading. However, it is certain that Diocles must have written "million", for we have some information on on estimates of distances between the celestial spheres made by his predecessors and contemporaries. The work ascribed to Hippolytus, *Refutation of all Heresies*, gives the distances in stades from the earth to the moon according to Aristarchus of Samos, Apollonius and Archimedes, and the distance from moon to sun, sun to Venus etc. according to Archimedes (IV 8-11, Wendland pp. 41-43). The distances from earth to moon range from about one million to about five million stades, and the distances between the other bodies from about 20 million to about 60 million stades. In particular, the distance of the sun from the earth (which is what Diocles is really interested in here) is, according to Archimedes, 55,806,195 stades (obtained by adding the distances earth-moon and moon-sun, Wendland p. 41, 13-15). It is typical of early Hellenistic astronomy to give the distances in stades, rather than in earth-radii, which became the norm after Hipparchus.

sṭādyūn This probably represents, not the nominative singular στάδιον, but the genitive plural σταδίων (which confirms that the original formulation was in myriads, see preceding note). The translator simply transliterated the word as it stood in his exemplar. In the Arabic translation of Ptolemy's *Planetary Hypotheses* the word is consistently rendered by "asṭādyā", which similarly, we may conjecture, represents στάδια in the original: see Goldstein's edition, e. g. p. 31, 3.

21 this second opinion The reference is, not to the 50 million stades as opposed to the 30 million, but to the opinion of the mathematical scientists (namely that the distances are of the order of 30-50 million stades) as opposed to that of those who scoff.

"fī-hā" refers to "al-sabīl" (which can be feminine).

time-measuring instruments which use the shadow "ālāti ʾl-sāʿāt" (literally "instruments of the hours") certainly represents ὡρολόγια, a standard term for "sundial". "allatī yustaʿmalu fī-hā al-ẓill" probably represents σκιοθηρικά or σκιακά. It is purely fortuitous that σκιοθηρικός is not attested before Strabo (*Geography* II 5 24, Aujac p. 107); for that very passage is a quotation from Eratosthenes: αὐτὸς δὲ διὰ τῶν σκιοθηρικῶν γνωμόνων ἀνευρεῖν (the distance from Rhodes to Alexandria), so the word was probably in use by 250 B. C. In fact Pliny (II 187, Beaujeu p. 82) attributes the invention of "horologium quod appellant sciothericon" to Anaximenes (sixth century B.C.), cf. Suidas (Adler I p. 536, s. v. γνώμων): ὅπερ ἐφεῦρεν Ἀναξίμανδρος καὶ ἔστησεν ἐπὶ τῶν σκιοθήρων. A σκιακὸν ὡρολόγιον is mentioned in an inscription of Pergamum, IGR 4 no. 293 col. I line 35, one of a series of decrees in honor of the well-known Diodorus Pasparos, and thus datable to soon after 130 B.C. (for the precise date of 125 B.C. for this inscription see Robert, *Opera Minora* I p. 156).

23 qiyās The sense is dubious. I have translated "analogy", taking it to refer to the analogy between position on the surface and position at the center, mentioned at the end of the sentence. However, it can equally well mean "example" or even "hypothesis". In texts translated from the Greek "qiyās" usually represents ἀναλογία, but it may also represent the simple λόγος. Thus ἐκ λόγου in Galen's *The Best Doctor is also a Philosopher* (*Scripta Minora* II p. 5, 16), which means simply "by reasoning", is rendered "mina ʾl-qiyās" in the Arabic version (Bachmann p. 20 line 80).

24 What could a sundial which uses the shadow but does not have a gnomon be? It is true that there is known from antiquity a type of sundial, with a spherical receiving surface, in which the gnomon is replaced by a hole cut in the roof of the instrument, so that the whole surface is in shadow, and the hour is marked by a ray of light. (On this "roofed spherical dial" see Gibbs, *Sundials* pp. 30-31 and nos. 9001G ff.; for a published example e. g. Diels, *Antike Technik* Pl. XI and pp. 25-26). But this is a trivial variation on the gnomon, and is open to exactly the same objection of

ignoring possible solar parallax as the gnomon itself (see note on 16-37, p. 144). The only way for this sentence to make sense is if it refers to the mirror sundial mentioned 15-16. But such a dial does *not* "use the shadow". It is possible that the mention of the shadow here is due to Diocles' using ὡρολόγια σκιακά as a general term for "sundials" (as opposed to e. g. waterclocks, which are also called ὡρολόγια: see LSJ s. v.), even for those sundials which use ray rather than shadow; or, less probably, to the translator's using the whole phrase "ālāt... al-ẓill" to render the simple ὡρολόγιον.

27 The translation of this sentence is merely a stop-gap, relying on extensive emendation of the manuscript text, which is probably irremediably corrupt.

29 It is hard to guess to what surfaces Diocles can be referring, especially since no treatise on sundials survives from antiquity. Of the actual surviving ancient sundials, the shadow-receiving surfaces are either plane, spherical or conical. One might conjecture that the conical surfaces are among those described as "very difficult to make", but in fact over a hundred examples survive from antiquity, more than of any other type (see Gibbs, *Sundials* pp. 34-42 and nos. 3001 to 3303). The invention of the type of sundial called "conus" (presumably with a conical surface) is ascribed by Vitruvius, IX 3. 1, to Dionysodorus, a contemporary of Diocles. In the same passage Vitruvius mentions several other types of sundial, not all of which can be identified with certainty.

33 quwan-humā One might perhaps interpret the ms. reading as a vulgar spelling of "aqwā-humā" ("the stronger of the two"), and translate: "the base of the stronger of the two in burning". But my emendation makes better sense.

34 foot The foot (ποῦς), though less common as a unit of measure in Hellenistic times than the cubit (cf. 31), occurs quite frequently. There is a contemporary example in an inscription from Lebadea concerning the building of a temple, SIG^3 972,106, μὴ ἔλαττον ἢ ἐπὶ δύο πόδας ἐκ τοῦ προσιόντος ἁρμοῦ. Cf. also [Heron], *Geometrica*, Heiberg p. 184.

aḥrā (more likely) Literally, "more fitting".

34–35 <u>sab^ca (seven)</u> An equally possible way of reading the ms. in both places is "tis^ca" ("nine"), but this is immaterial to the sense.

36–37 I suspect this passage of being an interpolation in the Arabic tradition, but cannot prove it. The suspicious features are (1) the alleged use of glass to make lamps; (2) the emphasis on the remoteness of the cities where the sacrifices take place; (3) the promise to do the same trick (which is not fulfilled in the work as we have it). However, there is nothing in the passage which is specifically un-Greek. There are many examples of the use of scientific principles to produce "miracles", especially in temples, in Heron's work on pneumatics, which certainly draws on earlier Hellenistic tradition (see e.g. Heron, *Pneumatics*, Schmidt I Props. 12, 17, 32, 38). Since the inside of Greek temples was dark, lamps were naturally a standard item of equipment in them. A good collection of references to lamps in temples in papyri and inscriptions is given by Robert, *Etudes Anatoliennes* p. 33 n. 1. However, the use of glass is odd. There is no mention of glass lamps before the fourth century A.D. (see the invaluable monograph of Trowbridge, *Ancient Glass*, pp. 190-91 for references), and the use seems characteristic of Byzantine times and Christian churches (ibid.). If our text is to make sense in context, it must refer to a *reflecting* surface. Glass mirrors were known in antiquity, but again the references are late. (The earliest is the elder Pliny, see Trowbridge pp. 184-86, Kisa, *Das Glas im Altertume* pp. 357-59; ibid. pp. 360-61, references to extant examples, the oldest allegedly from Ptolemaic Egypt). I suspect, however, that the writer of these lines was, confusedly, thinking of the burning-glass. This was known from at least the fifth century B.C. (Aristophanes, *Clouds* 768, with the scholia ad loc., quoted by Trowbridge pp. 178-79; see Kisa p. 357 for references to extant glass lenses from antiquity). If the reference is indeed to a burning-glass, it is so inapposite here that it is impossible to ascribe the passage to Diocles.

38–50 Prop. 1. Proof of the focal property of the parabola.

For other ancient and medieval proofs see Appendix B.

38 From now on I use the modern terms "parabola", "axis", etc., instead of the archaic "section of a right-angled cone", "line which bisects it", etc., which are used throughout by Diocles.

It is significant that the half-parameter is drawn as part of the figure and connected at B. I believe that it should be drawn as in Fig. 1, for it is perpendicular to BE, as is clear from 60; but it must be imagined as perpendicular to the plane of the parabola; it then represents the distance from the vertex of the parabola to the apex of the right-angled cone from which the parabola is generated. This corresponds exactly to the Archimedean nomenclature for the parameter, the "double of the distance to the axis" (see Introduction p. 6). In other words, BH represents a real line in three-dimensional space, and Diocles is using not only the pre-Apollonian nomenclature, but also the pre-Apollonian *definition* of the parabola. (For an unresolved difficulty here see Introduction p. 13 n. 5).

40 AB = BG I. e., the distance from the point where the tangent meets the axis (produced) to the vertex equals the distance from the vertex to the ordinate from the point of tangency. This is stated by Archimedes (in a more general form applicable to any diameter of the parabola and not merely the axis), *Quadrature of Parabola* II (Heiberg II p. 226), and said to be proven "in the elements of conics" (on the meaning of which see Introduction p. 5). It is proven by Apollonius, *Conics* I 33 and 35.

meets AZ beyond E This is far from obvious, and since it is stated without proof we may assume that it too was a theorem in "the elements of conics". It is a consequence of the theorem that the subnormal, GZ, is equal to the half-parameter (see next note). Since BE equals the half-parameter (39), GZ equals BE, and since the perpendicular from the tangent falls at G below B, Z must likewise fall below E.

41 GZ = BH I. e. the subnormal is constant and equal to the half-parameter. That this too was a theorem in the "elements of conics" is shown by its being assumed without proof here and by Archimedes, *On Floating Bodies* II 4 (Heiberg II p. 358; but Heiberg's figure is wrongly lettered, and some of his notes

incorrect; see rather Heath, *Archimedes* pp. 268-69). It can be proved simply, e.g. as follows (cf. Dijksterhuis, *Archimedes* p. 74):

$GZ.AG = \theta G^2$ (from the similar triangles θGZ, $AG\theta$)
$\theta G^2 = 2BH.BG$ (basic property of parabola, cf. 123, note)
but $AG = 2BG$ (see note on 55)
$\therefore GZ = BH$.

Curiously, Apollonius does not mention this property of the parabola in the elementary section of his *Conics* (Books I to IV). It can be derived only by combining the results of V 13 and V 27. This shows that there was more in the "elements of conics" than can be found in Apollonius' "elements".

42 GB = BA See 40.

44 triangle AθZ...Dθ = DZ Cf. Euclid III 31 (angle in semicircle).
The use of single letters (different from the letters marking the points of the figure) to denote angles is rare, and is probably another archaic feature of our text. It never occurs in the existing versions of Euclid's *Elements*. It is found, however, in Aristotle, e.g. in his proof that the base angles of an isosceles triangle are equal, *Prior Analytics* I 24, 41b5-22, and in the *Catoptrics* ascribed to Euclid *(Opera* VII p. 288 and *passim)* Both Aristotle and the *Catoptrics*, like Diocles, use "angle AB" to mean "angle A plus angle B".

A = PQ No use is made of this in the proof. Note however that in the proofs of this property in the Bobbio Mathematical Fragment and ibn al-Haytham (Appendix B), the equivalent statement is used, together with the equality of angles A and T, to prove the theorem.

45 fal-yujāz Here and in every other place in the work where "li-" plus the passive of the IVth form of a hollow verb occurs we find this strange form (instead of the expected "yujaz" etc.). See 66, 113, 131, 163, 237. I leave unanswered the question whether we should interpret this as an aberrant form of the jussive or an aberrant usage of the subjunctive (against the latter would count the regular use of the jussive in "li-yakun", which occurs many times in the work). See also note on 238.

The use of "jāza" IV here is standard. Cf. Nix p. 13 (equated with διάγειν) and e.g. ms. Marsh 667, 11^V, "wa-nujīzu ʿalā nuqṭa M kaṭṭan muwāziyan li-kaṭṭ DH ʿalayhi MN" = καὶ διὰ τοῦ M τῇ ΔΕ παράλληλος ἤχθω ἡ MN (Apollonius, *Conics* I 12, Heiberg I p. 44, 13-14).

51 brass (ṣufrin) One would expect "bronze", which was the material from which ancient mirrors were normally made. It is possible that the translator rendered χαλκός ("bronze") by "ṣufr": there seems to be no specific term for "bronze" in Arabic; the modern language uses "brūnz", which is simply a transliteration of the European term. However, there is no doubt that the usual meaning of "ṣufr" is "brass", as one would expect from the etymology (the root ṣfr means "yellow", which is applicable to brass but not to bronze). The only examples I have found of χαλκός being rendered by "ṣufr" are in Themistius, *De Anima* e.g. p. 99, 14 Heinze, ὥσπερ χαλκευτικὴ τοῦ χαλκοῦ = "miṯāl ḏālika anna ṣināʿati ʾl-ṣaffārīn ḵārija ʿan al-ṣufri" (Lyons p. 179, 13, cf. ibid. 9, 16; 179, 15). It is normally translated as "nuḥās" ("copper"): ibid. 46, 13; 99, 3; Heron, *Mechanics* II 1 (Nix p. 95, 11) "min nuḥāsin" = χαλκεύς (ibid. p. 272, 22).

If we were to assume that "brass" were the correct translation of what was in the Greek, the only word that it could represent would be ὀρείχαλκος. This in turn presents a further difficulty. It is most unlikely that all mentions of ὀρείχαλκος in extant texts refer to brass, particularly in earlier periods (it occurs already in Hesiod). But it is certain that in Roman times "orichalcum" normally meant an alloy of copper with a high percentage (about 20%) of zinc, i. e. what we should call brass. There is a large modern literature on the meaning of ὀρείχαλκος and brass manufacture in antiquity. The best discussion is by Caley, *Orichalcum*, especially Chs. III and VIII, on the origin and manufacture respectively. Also of some use is Forbes, *Ancient Technology* VIII pp. 272-86. On present evidence it is hard to be sure whether brass was produced in the Greek world during the earlier Hellenistic period, and, given the uncertainties of the translation, our text does not provide an answer. It is true that brass would be a suitable material for the kind of mirrors Diocles envisages, since it can be polished to a high shine, and can also easily be hammered to a desired shape. But, even assuming that ὀρείχαλκος stood in his text, he may not have meant "brass". For Ethel Eaton suggests that in earlier times ὀρείχαλκος meant "arsenical copper", which was used to plate metal objects with a very shiny surface. It would be especially appropriate for a mirror. In the fourth century B. C. Plato (*Critias* 116c1) mentions ὀρείχαλκος as a metal with a "fiery resplendence" (μαρμαρυγὰς ἔχοντι πυρώδεις), and in the third century Callimachus refers to a reflecting surface, perhaps a mirror, of that metal (*Hymns* V, 17-20, οὔσετε μηδὲ κάτοπτρον...οὔτ' ἐς ὀρείχαλκον μεγάλα θεὸς... ἔβλεψεν). There are three fourth-century Attic inscriptions mentioning the metal, which are of importance as attesting the real use (as opposed to literary allusion) of a substance with that name at a comparatively early period. These are IG II^2 1416 line 1, 1517 col. II line 83 and 1533 line 24. The last two are lists of dedications of objects, to Artemis Brauronia and Asclepius respectively. The dedication to Asclepius is of "two chains" [i. e. ornamental chains, such as necklaces] "one of copper, the other of iron, the first being of ὀρείχαλκος". This provides considerable support for Eaton's identification.

53 - 59 The operations described will indeed cause the parabola's focus, D, to move on a circle (or rather semi-circle), but the result will not be a surface from which the sun's rays are reflected to a circle. For the rays will no longer, as in the case of the parabolic mirror (51-52) all be in a plane containing a parabola. See Appendix D, where it is shown that most rays are *not* reflected through the locus of D. It is not easy to determine what curve in the plane of that locus the rays are reflected through, or indeed whether there exists a curve in that or any other plane through which all rays are reflected from the surface. It was certainly far beyond the capabilities of ancient mathematics to answer such a question. However, rather than attributing to Diocles the simpleton's error of thinking that by moving the focus in a circle he ensured that the rays were reflected to a circle. I believe that he is taking into account the motion of the sun during the day. For if the plane perpendicular to the plane of the paper and passing through AZ is set up in the plane of the equator, then (at least at the equinoxes), at any moment of the day the plane perpendicular to the plane of the equator passing through the sun and the center of the

semi-circle which is the locus of D will intersect the surface formed by the rotation of FBM in a parabola, and thus the rays in that plane will all be reflected to a point on that semi-circle. See note on 16-37, pp. 143-4, where the possible application to a sundial is discussed.

53 which is like a chord of section FBM Since there exists no Greek term corresponding to "chord" ("watar") (the nearest is αἱ ἐν κύκλῳ εὐθεῖαι, "straight lines in a circle"), this otiose statement is probably an interpolation in the Arabic.

56 radius Literally, "the line drawn from the center to the circumference", almost a word for word translation of the standard Greek term, ἡ ἐκ τοῦ κέντρου.

57 perform those other operations I. e., construct a parabola etc. (see Prop. 4 for the method Diocles uses to construct a parabola from a given focal distance).

bi-aʿyāni-hā The plural is used because of the previous plural "al-ašyā'". For the expression cf. Georr, *Catégories* p. 235 no. 179.

59 some plane Any plane through FE will do, since it is the axis of revolution.

60-62 The motion described will indeed generate a surface from which the rays will be reflected to the circumference of a circle (if the axis of revolution coincides with the direction of the sun). However, the description of the motion is incomplete, since we are not told how the line BM behaves during its revolution. Does it, for instance, remain parallel to itself? This would produce a self-intersecting surface. It seems far more likely that Diocles envisaged it turning uniformly about the circle BLN, so that the parabolic arc BM always lies in the same plane as the radius of circle BLN on which B lies. The motion of BM round the arbitrary curve BLN of 63-67 could be governed in a similar way, by designating some point within the curve as its "center". The resultant surface (in the case of a circle) would be a modified paraboloid, similar to that produced by rotating a parabola about a diameter other than the axis.

60 I have drawn both BH and circle BLN, according to the ancient convention, as if they lay in the plane of the original parabola, but in fact both lie in planes at right angles to it. For the location of BH see note on 38.

65 size and shape literally, "amount" ("miqdār").

66 wal-yudār For the form see note on 45.

which is tangent to the curve... BLN[1] Perhaps this clause should be omitted as a (scribal) repetition of the clause just below ("which is tangent to curve BLN"). Certainly BM is not "tangent" to BLN in the mathematical sense (whereas ZB is). One can say loosely that it "touches" it. Or did Diocles mean that the tangent to BM at B lies in the plane of the curve BLN?

at right angles I. e. B and every point on BM move on planes at right angles to the "established plane".

After "the established plane" the ms. adds "which is the plane which passes through line BH". I have excised this phrase as a stupid gloss interpolated in the Arabic tradition (perhaps due to a misunderstanding of the conventions of the figure, see note on 60). BH (see note on 38) lies, not in the plane of the paper, but in a plane at right angles to it.

fixed in its original position Not stationary (hence "lāzib", not "ṯābit"), but always perpendicular to the plane of curve BLN.

68–77 Prop. 2. In a spherical mirror all rays parallel to a given radius are reflected through the half of that radius nearer to the surface.

A similar proposition is proved in the Bobbio Mathematical Fragment, *Mathematici Graeci Minores* pp. 88-89. Euclid, *Catoptrics* Prop. 30, first part (Heiberg p. 340), "proves" that all rays will be reflected to points on the radius between the center and the circumference.

68 some line This should rather be "some radius", as is obvious, since the center B lies on the line.

69 $\widehat{BLM} = \widehat{BLG}$ For LM = LG (construction).

72 Euclid III 7.

73 LE > EG Euclid III 7

74 $\widehat{CP} = \widehat{ROQ}$ For GL = GZ (construction).

75 $\widehat{QO} = \hat{C}$ By construction (70).

BX > XA Proof: since $\hat{D} = \hat{R}$, BX = GX. But GX > XA (cf. the proposition of Euclid utilized in 72 and 73), therefore BX > XA. This is a rather large jump over intermediate steps, and perhaps we should posit a lacuna in the manuscript (easily explained by haplography, the eye of the scribe going from one mention of BX to the next). However, omission of obvious intermediate steps is characteristic of Archimedes, and Simplicius *(Comm. on Physics* A 2, Diels p. 60, 29-30) remarks that it was "the ancient manner" to set out proofs summarily, so we may regard this as a sign of archaism in Diocles. For an even more drastic example see 83.

76 pass between points A and H For $\hat{C} = \widehat{QO}$ (75), hence SG is reflected as GX, and X lies between A and H.

77 For $\hat{C} > \hat{Q}$ (74), therefore E, where the reflection of FL cuts AB, is nearer to A than is X, where the reflection of SG cuts AB.

78 – 96 Prop. 3. In a spherical mirror all rays parallel to a given radius striking an arc of the mirror 30° either side of that radius are reflected through a section less than $^1/_{12}$th of the radius.

78 the established plane The same expression occurs at 53 and 66, but I do not see the point of it here.

79 DA, AF . . . a sixth of the circumference For HB = $^1/_2$ radius = $^1/_2$ DB, so DBH is a triangle right-angled at H, with its hypotenuse equal to twice one of its sides. Therefore $\widehat{DBH} = 60°$.

81 equal angles with arc ΨAF The use of an "angle" between a straight line and an arc of a circle is probably an archaism.

An early example is Aristotle, *Prior Analytics* I 24, 41b. See also [Euclid], *Catoptrics* Prop. 30 (Heiberg p. 341). Euclid defines such angles in the *Elements* (III Def. 7), but never uses them. Hence Heath (*Euclid* II p. 4) infers that the definition is taken over from earlier textbooks. It is interesting to find such angles also in the Bobbio Fragment's proof of the focal property of the parabola (see Appendix B(i) p. 203).

<u>as we showed above</u> Prop. 2.

82 <u>let XG... equal to $\hat{S}$</u> This *defines* the position of K.

<u>$\hat{S}$ is $1/3$ of a right angle</u> For $\widehat{AG} = 1/2\ \widehat{AD} = 1/12$th of circumference (79).

83 It is far from obvious that $\hat{R}$ is a right angle. Of the various ways to prove it perhaps the simplest is as follows (I use Diocles' terminology for the size of angles). Since TG is parallel to BA, $\widehat{TGB} = \widehat{ABG} = 1/3$ of a right angle. Therefore $\widehat{DGT} = 2/3$ of a right angle $= \widehat{AGX}$. Therefore $\widehat{TGX} = 2$ right angles minus $(\widehat{DGT} + \widehat{AGX}) = 2/3$ of a right angle. Therefore $\widehat{TGK} = \widehat{TGX} + \widehat{XGK} = (2/3 + 1/3)$ of a right angle = one right angle. Since TG is parallel to BA, $\widehat{GKB}$ is also a right angle. Cf. note on 75, p. 156.

84 HA and HB are each equal to half the radius (by construction). GK is also equal to half the radius, since $GK = 1/2$ GN, and GN is the side of the hexagon inscribed in the circle, and hence equals the radius. (Alternatively one can show that the triangles DHA and GKB are congruent).

85 $\hat{S}$ equals $\hat{E}$ because both are $1/3$ of a right angle. For $\hat{S}$ see 82. $\hat{E} = \widehat{TGK} - (\widehat{TGB} + \widehat{KGX}) = 90° - (30° + 30°) = 1/3$ of a right angle.

86 $BZ^2 = 4/3\ BH^2$ because in the 30°-60°-90° triangle BZH, $BZ = 2ZH$, therefore $BZ^2 = 4ZH^2$. And $BH^2 = BZ^2 - ZH^2 = BZ^2 - 1/4\ BZ^2 = 3/4\ BZ^2$.

87 Diocles wishes to evaluate the length of the section XH, through which all rays reflected from arc GAN must pass, in terms of the radius. Since BX = BZ (85), this can be done

via the 30°-60°-90° triangle BZH, namely $\frac{XH}{BH} = \frac{BZ - BH}{BH}$. In modern terms, this is sec 30° - 1. However, in Diocles' time trigonometry had not yet been developed (on this see Toomer, "Chord Table of Hipparchus" pp. 16-23), so precise evaluation was not possible. Instead, Diocles provides an approximation, using an inequality, exactly in the manner of Aristarchus and Archimedes. It is not obvious how he arrived at this inequality. Perhaps the simplest way is as follows.

A lower bound for $\sqrt{3}$ is given by

$$3 = \frac{147}{49} > \frac{144}{49} \quad \therefore \sqrt{3} > \frac{12}{7}.$$

$$\frac{BX^2}{BH^2} = \frac{4}{3} \qquad \therefore \frac{BX}{BH} = \frac{2}{\sqrt{3}}.$$

$$\frac{HX}{BH} = \frac{BX - BH}{BH} = \frac{2 - \sqrt{3}}{\sqrt{3}} < \frac{14 - 12}{12} = \frac{1}{6}.$$

$\therefore$ BH > 6HX.

(Similarly, by establishing the upper bound, $\sqrt{3} < \frac{7}{4}$, one can show that BH < 7HX).

It is conceivable that Diocles simply used Archimedes' bounds for $\sqrt{3}$ (*Measurement of a Circle*, Heiberg I pp. 236-40). These are

$$\frac{1351}{780} > \sqrt{3} > \frac{265}{153},$$

which lead immediately, by the same method as above, to

$$6\frac{97}{209} > \frac{BH}{XH} > 6\frac{19}{41}.$$

However, if Diocles did use these numbers, it is hard to see why he did not announce the more accurate result which they supply.

88 This follows from 77.

89 <u>the section beyond X</u> Literally, "the exterior (k̲ārij) of point X".

90 The nearer. . . reflection is to H This also follows from 77.

97-111 Prop. 4. Construction of the parabola from given focal distance (use of the focus-directrix property to construct the parabola).

97 center of the surface I. e., the vertex of the parabola.

98 AK must be at right angles to AB, though this is not stated.

99 EF is equal to FK For AK = 2AB = 2AF, and AF = EF.

101 it cuts GM For GM > GA.

102 N Here and in what follows the letter denoting "nūn" often looks more like "rā" or "zāy" in the ms. I have preferred to interpret it as a badly-written "nūn" rather than make an explicit correction every time. However, in Fig. 5 the point corresponding to Z of Fig. 4 is unambiguously N ("nūn") in the ms. On the other hand, the positions of Z and N as I have marked them in Fig. 4 are guaranteed by the order of the Greek alphabet. But I suspect that they became interchanged at some point in the transmission (probably in the Greek, by rotation of the letters through 90°).

105 it passes through R For KA = 2AF, AR = 2AB.

106 LD = NA and MG = θA By construction (see 102, 101).

107 I. e. it is true for each of the points K, N, θ, B that its distance from A, the focus, is equal to its perpendicular distance from line SR. This is the focus-directrix property of the parabola.

LQ = NO and MC = θP because they are parallel lines drawn between parallel lines.

108 we shall prove subsequently In Prop. 5.

109 Here and elsewhere (195, 198, 219) Diocles refers to what is clearly a flexible ruler, which can be bent to help draw a continuous curve through a number of points. "misṭara" can represent only κανών, and the only other reference I

know from antiquity to a flexible κανών is the enigmatic "leaden rule" (μολίβδινος κανών), mentioned by Aristotle, *Nicomachean Ethics* V 10, 1137b30. From that passage it appears that this instrument, which was used "in the Lesbian method of building", could be bent to the required shape for the stone (πρὸς τὸ σχῆμα τοῦ λίθου μετακινεῖται). The material of Diocles' instrument, however, appears, from this passage alone, to be horn. The Arabic is "min qurūnin", literally, "of horns". One would expect the singular, "min qarnin", but, unless there is some corruption, I suppose a misunderstanding by the translator of some adjective such as κεράτινος. Horn was a very common material in antiquity, and its flexibility would make it suitable for the use envisaged here. The closest parallel I can find is its use for bows (for examples see LSJ s. vv. κέρας III 1, κερόδετος, κερουλκός). A surviving example of an artifact made out of horn because of its flexibility are the strigils, if they are correctly identified as such, found at Balabish in Egypt (New Kingdom): see Wainwright, *Balabish* p. 13 and Pl. XII 8.

110 <u>template</u> The Arabic is "qālab", meaning "mold" or "last".

112–24 Prop. 5. Proof that a curve produced by the focus-directrix construction is indeed a parabola.

On the significance of the fact that Diocles feels it necessary to prove this see Introduction p. 17.

113 <u>MG - MN = BE</u> Literally, "line MG exceeds line BE by line MN", i. e. this is a way of *defining* point N. There is a similar phrase at the end of 114.

114 <u>what was stated</u> At 99.

EN equals NM because ΔENM ||| ΔEFK ||| ΔRAK, and RA = AK.

115 <u>QA = MG</u> For QA = AB + BQ, and MG = AB + BG (114).

116 <u>since $\hat{G}$ is right</u> I. e., in triangle AGM, AM > MG.

118 <u>we have shown</u> At 115.

$4(AB.BG) + GA^2 = AQ^2$ This is a direct application of Euclid II 8, which states (in Heath's translation, I p. 389) "if a straight line be cut at random, four times the rectangle contained by the whole and one of the segments together with the square on the remaining segment is equal to the square described on the whole and the aforesaid segment as on one straight line". In the present case, AB is a straight line cut at G, therefore

$$4(AB.BG) + GA^2 = (AB + BG)^2.$$

Since QB = BG, AB + BG = AB + BQ = AQ, so Diocles' statement follows.

It is interesting that Heath remarks (ibid. p. 200) that the proposition is "of decided use in proving the fundamental property of a parabola". I presume he means that if a parabola is *defined* by the focus-directrix property, one can prove the relationship between ordinate and abscissa, $y^2 = px$, where p is four times the focal distance, using Euclid II 8. This is precisely what Diocles does, the only difference being that for him the *defining* property is the ordinate-abscissa relationship (cf. Introduction p. 6). See further ibn al-Haytham's procedure, Appendix B(ii).

123 I. e., the defining property of the parabola, for Diocles, is the constant relationship between the square on the ordinate and the abscissa of the diameter (see Introduction p. 6). Since the parameter T is 4AB, he incidentally proves that the point A is indeed the focus (in the sense of the point where the reflected rays all meet (cf. 9).

$AK^2 = AB.T$ For $AK^2 = (2AB)^2 = 4AB^2 = AB.4AB = AB.T$.

125-35 Prop. 6. "Proof" that equal sections of a straight line are seen under unequal angles by an observer situated outside the line.

That this pròposition is a spurious addition is clear from several considerations. It is utterly trivial, being a mere variant on Euclid, *Optics* Prop. 4 (the proposition, not the proof). It appears completely alien from the rest of the work. The "proof" assumes known what is to be proven (128), employs a mechanical method alien to Greek geometry (129-30) and contains a gross fallacy (133). I have no doubt that it is an interpolation in the Arabic transmission, but can suggest no plausible motive for the interpolation.

125 are subtended by unequal (angles) Literally, "are seen as unequal".

We must prove that Perhaps an indication that only the statement of the proposition, and not the proof, is by Diocles. But I cannot think of a reason why Diocles should even have enuciated the proposition.

127 li-k̲uṭūṭin One would expect "ilā k̲uṭūṭin". Of the possible ways of reading the ms., "li-" seems preferable to "bi-".

128 The statement is true, but precisely what is supposed to be proved.

130 Refer to Fig. 6b.

131 in the preceding section At 128.

133 PX is subtended by $\widehat{HBD}$ This (like the corresponding statements in 133 and 134) is of course false. The writer makes the following assumption: in Fig. VI $\frac{\alpha}{\beta} = \frac{A}{B}$. But in fact $\frac{\alpha}{\beta} < \frac{A}{B}$, and this inequality, which is equivalent to the statement that, if $\alpha > \beta$, $\frac{\alpha}{\beta} < \frac{\tan \alpha}{\tan \beta}$, was well known in Greek mathematics before Diocles. It is used e. g. by Aristarchus of Samos (Heath pp. 376-78) and by Archimedes (*Sandreckoner*, Heiberg II p. 232), and was of fundamental importance in evaluating triangles before the development of trigonometry, as I shall show in detail elsewhere. It is inconceivable that any competent mathematician could have committed such a flagrant error as we find here.

136 – 49 Prop. 7. Reduction of Archimedes' problem "how to cut a given sphere by a plane so that the two segments are in a given ratio" to conditions affording a solution by means of conic sections.

For Eutocius' version of this and Prop. 8 see Appendix A(i).

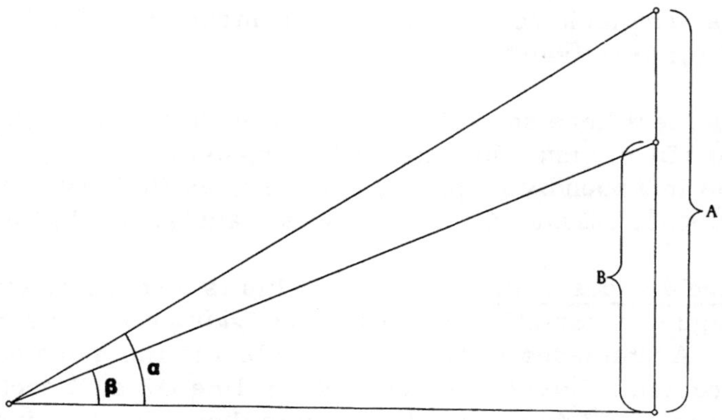

Fig. VI

136–37 This is proved by Archimedes in *Sphere and Cylinder* II 2 (Heiberg I pp. 174-78).

137 perpendicular. i. e. height (ʿumūd. . . allaḏī huwa irtifāʿu-hā) Eutocius' version of Diocles (Heiberg III p. 160, 12) has simply κάθετος ("perpendicular"). Archimedes, on the other hand, uses ὕψος ("height") (Heiberg I p. 174, 8). So what we have here is not a translator's gloss, but was already in the Greek, though it seems impossible to decide whether it was written by Diocles himself or by a later glossator who compared Archimedes' text.

139 by a similar construction Eutocius (p. 160, 17-18) gives explicitly $(EB + BZ) : ZB = \Theta Z : ZA$.

140 ḵaṭṭ ZH. . . wa-sahmuhu This essential clause, missing in the ms., has been restored from Eutocius (p. 160, 20-22). The omission is due to haplography.

141 This is the subject of *Sphere and Cylinder* II 4 (Heiberg I pp. 186-94).

142–43 Archimedes, Heiberg I p. 188, 1-6.

142 and it is equal to the ratio of ΘZ to ZH This sentence is missing in Eutocius' version (Heiberg p. 160, 28), though it is essential to the sequence of ideas. Misled by this,

Heiberg took τοῦτο in the next sentence to refer back, instead of forward (see next note).

143 The text here shows that the text of Eutocius' mss., ἀπεδείχθη οἱ κῶνοι, must be emended to ἀπεδείχθη <ὅτι> οἱ κῶνοι (as in Appendix A, p. 180), and not, as Heiberg, ἀπεδείχθη. οἱ <δὲ> κῶνοι. Cf. Moerbeke's translation, "quod coni".

149 <u>Archimedes... did not solve</u> This is a correct, though vague, description of *Sphere and Cylinder* II 4, as we have it. Archimedes reduces the problem to the more general problem, "divide a given straight line DZ at X such that the ratio of XZ to a given line equals the ratio of a given area to DX squared" (Heiberg I p. 190, 22-25). Having shown that the problem of dividing the sphere in a given ratio is a particular case of the above, he concludes (Heiberg p. 192, 5-6) "the analysis and synthesis of both [the general problem and the particular case] will be given at the end". This promise is not fulfilled in the existing text of *Sphere and Cylinder*, and, if it ever was, Archimedes' solution had already been lost soon after his death, since both Dionysodorus (see Introduction p. 2) and Diocles provided solutions, not indeed of the general problem propounded by Archimedes, but of the original problem of dividing the sphere. We owe our knowledge of these (until the discovery of the present text) to Eutocius' commentary on the *Sphere and Cylinder*. Eutocius also gives a solution of the general problem, which he ascribes, plausibly, to Archimedes himself. See Heiberg III pp. 130-76; Heath, *Archimedes*, pp. 65-79.

150–74 Prop. 8. Solution of the problem defined in Prop. 7. See Appendix C for a treatment of the problem in modern terms.

It is noteworthy that Diocles changes some of the letters from those used in the definition of the problem in Prop. 7 (e. g. the point at which AB is to be cut, which was Z in Prop. 7, is G in Prop. 8). To complicate matters further, Eutocius has lettered the figure in yet another way.

152 Points Z and K must be further defined by saying that they lie on the extensions of ΘB and HA respectively. The corresponding points Λ and Θ in Eutocius (Heiberg III p. 162, 28) are similarly ill-defined.

153 $DA : AG = \theta B : BG$ (151)
$\theta B : BG = KA : AG$ (similar triangles).
Notice how Eutocius gives precisely the content of this note in his version.

154 It seems likely that this is an interpolation in the Arabic transmission. It has no counterpart in Eutocius' version (though this is far from conclusive). More important, though the statement is true, it has no consequences, and interrupts the flow of the proof (see next note). Furthermore, what I have translated as "rectangle Aθ" is literally "the figure Aθ having equal sides", i. e. "parallelogram Aθ"; but Aθ is given from its sides only if it is a rectangle. The last point, however, is not conclusive: in Archimedes' *Method* (Heiberg II p. 418, 23, al.) παραλληλόγραμμον has the meaning "square". Although this is almost certainly not Archimedes' own usage (cf. Introduction p. 4 n. 3), at least it provides a Greek parallel for the usage here.

155 $DG : GE$ is given (150)
and $DG = DA + AG = KA + AG$ (153),
$GE = GB + BE = GB + BZ$ (153).
Therefore $(KA + AG) : (BZ + BG)$ is given.

157 By similar triangles

$$\frac{AH}{BZ} = \frac{AG}{BG} \text{ and } \frac{KA}{\theta B} = \frac{AG}{BG}.$$

By addition of ratios

$$\frac{AH + AG}{BZ + BG} = \frac{AG}{BG} = \frac{KA + AG}{\theta B + BG}.$$

159 See 156.

160 <u>L and M are given</u> For $AL = BM = AH$ = the given line (152).

161 <u>DG. GE = LG. GM</u> For $DG = KA + AG$, $GE = ZB + BG$,
$LG = HA + AG$, $GM = \theta B + GB$,
and (158) $(HA + AG)(\theta B + BG) = (KA + AG)(ZB + BG)$.

162 Since $LG. MG = DG. GE$,

$$\frac{LG.MG}{(ZB+BG)^2} = \frac{DG.GE}{(ZB+BG)^2} = \frac{(KA+AG)(ZB+BG)}{(ZB+BG)^2},$$

which is given (156).

165 By the similar triangles LBQ, GBP, MBR (cf. note on 167).

166 $$\frac{LG^2}{LG.GM} = \frac{LG}{GM} = \frac{QP^2}{QP.PR},$$

therefore

$$\frac{LG^2}{QP^2} = \frac{LG.GM}{QP.PR}.$$

167 $$\frac{LG}{QP} = \frac{LB-GB}{QB-PB},$$

but

$$\frac{LB}{QB} = \frac{GB}{PB},$$

therefore

$$\frac{LB-GB}{QB-PB} = \frac{GB}{PB},$$

therefore

$$\frac{LG^2}{QP^2} = \frac{GB^2}{PB^2}.$$

168 ZB + BG = SG + GP = SP.

169 So (QP.PR) : SP^2 is given This is missing in the text, but is absolutely necessary. It is restored from Eutocius' version (Heiberg III p. 168, 3-4).

170 S lies on... an ellipse given in position I. e., to use modern terminology, if y is an ordinate, x_1 and x_2 the (non-overlapping) segments cut off on a diameter by that ordinate, the equation $\frac{y^2}{x_1x_2}$ = constant *defines* an ellipse.
In this case QR is the diameter, SP the ordinate. It is not surprising to find this relationship used as the definition of an ellipse: Archimedes takes it for granted, e. g. *Conoids and Spheroids* VIII (Heiberg I p. 294, 22-26). What is re-

markable is, firstly, that the ordinate is not at right angles to the diameter, i. e. we have the equation applied in oblique conjugation (whereas Archimedes normally uses it in orthogonal conjugation, i. e. applied only to the principal axes, not to any diameter); and, secondly, the name given to the ellipse, Arabic "qaṭ' nāqiṣ", which can represent only ἔλλειψις (instead of "section of an acute-angled cone"). Both features point *prima facie* to use of Apollonius' *Conics* (the above property is proven *Conics* I 21). It is also noteworthy that we have in 169 exactly the elements required by Apollonius (Conics I 58) for constructing an ellipse, namely (1) the diameter (RQ); (2) the parameter, which is simply the line length which bears the same ratio to the diameter as the square on the ordinates bears to the rectangle contained by the segments of the diameter, i. e. here the parameter equals $\frac{SG^2}{QP.PR} \cdot QR$; and (3) the angle between the ordinates and the diameter ($\widehat{SPR}$). However, it is evident from *Conoids and Spheroids* XXVI and XXVIII that Archimedes was well aware of the property of the ellipse (and the corresponding property of the hyperbola) in oblique conjugation, and the argument from nomenclature is not conclusive either. See Introduction pp. 9-15.

<u>rectangle Aθ is equal to rectangle SH</u> Rectangle NG = rectangle Gθ (complements of parallelograms about the diameter, Euclid I 43). Add rectangle AO to both.

171 Here again (cf. note on 170, pp. 166-7) the terminology suggests, at first sight, that Diocles used Apollonius' *Conics*. The term for "hyperbola" is "qaṭ' zā'id", which is the standard Arabic for ὑπερβολή (and not "section of an obtuse-angled cone"). Furthermore the expression "al-ḵaṭṭayni 'llaḏayni lā yalqayāni-hu" must represent αἱ ἀσύμπτωτοι (the Arabic translation of Apollonius' *Conics* uses the similar phrase "al-ḵaṭṭāni 'llaḏāni lā yaqa'āni 'alā 'l-qaṭ'i", see Nix p. 12); but ἀσύμπτωτοι does not occur in the sense of "asymptotes" in extant Greek literature before Apollonius (Archimedes uses αἱ ἔγγιστα τᾶς τομᾶς or similar phrases: see e. g. *Conoids and Spheroids* Introduction, Heiberg I p. 248, 24-25). However, the second case is less cogent than the first, since the word seems to have existed in other applications (e. g. in the

fourth century B.C., Autolycus, *Spherics* 8, Mogenet p. 207, 5 and 15, if the text we have is indeed that written by Autolycus), and could have been applied to the hyperbolic asymptotes (knowledge of which belongs to the earliest developments in the theory of conics, see below) by anyone.

Nevertheless, though the form may be reminiscent of Apollonius, the content is not. Diocles assumes that, given two straight lines meeting at right angles, the locus of points the product of whose vertical distances from the two lines is constant is a hyperbola to which the two lines are asymptotes. This is indeed proven by Apollonius, *Conics* II 12 (in the more general form where the asymptotes form any angle whatever, and the distances are taken along lines parallel to the asymptotes). However, it occurs in precisely the same form as in Diocles in the earliest known problem involving conic sections, the method of finding two mean proportionals discovered by Menaechmus (fourth century B.C.). It is true that we have this only in the version of Eutocius (on Archimedes' *Sphere and Cylinder* Heiberg III pp. 78-80), but this property of the rectangular hyperbola is essential to the solution, and must have been known to Menaechmus (it was probably discovered by him). It is also used in the solution (of the same problem which Diocles discussed here) which is given by Eutocius (Heiberg III p. 134, 18-24) and plausibly ascribed by him to Archimedes himself; and again in the solution of the same problem given by Diocles' contemporary Dionysodorus (ibid. p. 154, 19-21).

174 <u>tarkīb</u> This is the standard translation of σύνθεσις, e.g. Apollonius, *Conics* II 49 (Heiberg I p. 286, 24), ἡ δὲ σύνθεσις ἡ αὐτὴ τῇ πρὸ αὐτοῦ, is translated (ms. Marsh 667, 45^{v}, 5-6) "wa-tarkīb <u>d</u>ālika yakūnu ʿalā miṯli mā qaddamnā bayāna-hu". See also ibn al-Haytham, *Majmūʿ*, third treatise, p. 4, 19.

175-85 Prop. 9. To construct a length equal to a given length plus a prescribed fraction of it.

This is certainly a spurious addition. Although mathematically correct, it is an utterly trivial variation on Euclid VI 9; as such it has no place in a serious work of higher mathematics. The only possible connection that it might have to the rest of the work is as an (unnecessary) explanation of 223, where one has to find a line (EN) which is in a given ratio to a given line (DE).

175 The last three words of the sentence seem hopelessly corrupt. One could emend to make sense, e.g. "t̲amāniyat asbāʿi ʾl-k̲aṭṭi ʾl-mawḍūʿi", "eight sevenths of the posited line", but the emendation is not very plausible.

180 as in the diagram An alternative translation is "as we have described", referring to 176 (but there it is DA that is perpendicular to DE).

186-207 Prop. 10. To construct a cube twice a given cube.

This and all subsequent propositions are related to the famous problem of "doubling the cube" (sometimes known as "the Delian problem", because of the story that an oracle of Apollo required the Delians, in order to be relieved of a plague, to construct an altar twice the size of an existing one). The problem goes back to the fifth century, for Hippocrates of Chios (late fifth century) reduced it to the problem of finding two mean proportionals between two lines one of which was double the other (Eutocius on Archimedes' *Sphere and Cylinder*, Heiberg III p. 88, 18-21). All ancient solutions are in fact solutions of the latter problem. For an account of the many solutions known from antiquity see Heath, HGM I pp. 244-70. Most of these (including both of Diocles' solutions) come from the commentary of Eutocius on Prop. 1 of Book II of Archimedes' *Sphere and Cylinder*, in which Archimedes takes it for granted that one can find two mean proportionals. Several of the solutions involve conic sections (as one would expect for a problem which, in effect, requires the solution of a cubic equation).

The present solution, involving the intersection of two parabolas, one of which has a focal distance (or parameter) twice the other's, is given by Eutocius (Heiberg III pp. 82-84; here Appendix A(ii)). However, Eutocius does not mention the author. It has been commonly assumed that it is due to Menaechmus, on the not very cogent grounds that in the text of Eutocius it follows a solution by Menaechmus, being introduced by the remark ἄλλως ("another way"). This erroneous attribution goes back a long way: it is found in Molther's *Problema Deliacum* of 1619, p. 23, which I quote because of the work's rarity: "Menechmus...hic vel Parabolae

& Hyperbolae vel duarum Parabolarum haud commodas descriptiones requirebat". Modern authors, to cite a few among many, include Montucla, *Histoire des Mathématiques* I p. 177, Schmidt, "Fragmente des Menaechmus" p. 77 and Heath, HGM I pp. 254-55. Since we now find the solution in Diocles' work, and since Eutocius provides two other excerpts from Diocles, it is certain that Diocles is his source for this too. It is true that Eutocius completely recast the proof, putting it in the "classical" form of analysis and synthesis (where Diocles gives only the "synthetic" form). But he does exactly the same for Prop. 8 (see Appendix A(i)), where Diocles gives only the "analytic" form. Eutocius also omits all reference to the method of generating the parabola from focus and directrix, preferring instead to specify the parameters. This is explained by his pedantic desire to present all theorems of conics in the "classical" version of Apollonius' (cf. his insertion of anachronistic references to Apollonius' *Conics* in his version of Diocles' Prop. 8, Heiberg p. 168, 11-12; p. 170, 16-17 and 22-23). Unfortunately for him Apollonius makes no mention of the focus-directrix property in the *Conics*.

191 k̲atīratan I am far from confident that this emendation is correct (it is rather far from the manuscript, which appears to have "mā kat̲ra": perhaps one could read "bi-kat̲rati", "in abundance"); but I am sure that I have rendered the sense intended by Diocles. It is irrelevant that points L, M, O, P, Q are neither "many" nor "close to each other", since they are only *exempli gratia*.

193–94 I. e., D is the focus, and the directrix is a line through G parallel to EZ. For this method of drawing the parabola see Prop. 4.

194 li-k̲aṭṭi "li" is not necessary, cf. 193 "qaṭʿi-hā k̲aṭṭi", but is permissible, cf. e. g. Apollonius, *Conics* I, ὅροι (Nix p. 5 [Arabic] lines 21-22) "wa-kāna kullu wāḥidin min-humā qātiʿan liʾl-k̲uṭūṭi".

195 the curved ruler See note on 109.

in this way This use of "jiha" as the equivalent of τρόπος, not recognized in most dictionaries, occurs not infrequently in mathematical texts, e.g. Diophantus, ms. Meshhed, Shrine Library 295, p. 14, 16 "bi-jiha ukrā" (= ἑτέρῳ τρόπῳ), ibid. 16, 18; 61, 4; al.; cf. Nicomachus, ed. Kutsch p. 19, 20-21 (cf. index p. 276 s.v.); cf. Galen, *Compendium Timaei* p. 9 [Arabic] line 10.

196–97 Here the focus is E, and the directrix a line through Z parallel to GD. Cf. note on 193-94.

200 DN Literally, "the line drawn from D to N" (because it is not actually drawn in the figure).

GL = DN by construction (193).

$4LH.HD + DL^2 = DL^2 + LN^2$ See note on 118, p. 161. We have exactly the same situation here as in 118-20, but Diocles now omits some of the steps in the proof. The full sequence is:

$4LH.HD + DL^2 = (LH + HD)^2$.
$LH + HD = LH + HG = LG = DN$ (193),
$4LH.HD + DL^2 = DN^2 = DL^2 + LN^2$ ($\widehat{DLN}$ a right angle).

201 $4LH.HD = A.HL$ For $HD = \frac{1}{4}A$ (188).

205 mean proportionals Literally, "in continuous proportion". "mutawāliyāni ʿalā nisba" = δύο μέσαι κατὰ τὸ συνεχὲς ἀνάλογον, as Archimedes, Heiberg I p. 198, 12.

For the equivalence between finding two mean proportionals and doubling the cube see the introductory note to this proposition, p. 169. Since

$$\frac{A}{NL} = \frac{NL}{LH} = \frac{LH}{B},$$

$$LH = \frac{NL^2}{A} = \frac{A.B}{NL},$$

$$\therefore NL^3 = A^2.B = \frac{A^3.B}{A},$$

$$\text{and } \frac{A^3}{NL^3} = \frac{A}{B}.$$

207 obvious that lines θCNR, KNS are parabolas Proved Prop. 5.

208–13 Prop. 11. Method of finding two mean proportionals in a circle. For Eutocius' version of this and Props. 12 and 13 see Appendix A(iii).

210 AΘ : ΘZ = ΘZ : ΘB For triangle AZB is right-angled at Z (angle in a semi-circle).

214-20 Prop. 12. Construction of a curve in the circle to solve the problem of finding two mean proportionals.

216 rābiʿ I know of no other example of this word with the meaning "quadrant". Possibly one should emend to "rubʿ", the normal term (see e. g. Nix, *Apollonius* p. 13).

218 in the preceding proposition See Prop. 11.

219 by means of the curved ruler See note on 109. Eutocius, in his version of this passage, seems to envisage joining a large number of points with a *straight* ruler, forming a series of very short straight lines instead of a smooth curve: κάνονος παραθέσει ἐπιζεύξαντες εὐθείας (Heiberg III p. 168, 10-11).

The line BRQPD is (part of) the curve known in modern times as the "cissoid". This name (κισσοειδής or κιττοειδής) was applied in antiquity to a certain type or class of curve, but it seems most unlikely that it was applied to the curve generated here by Diocles (for my reasons for saying this, and for an account of the naming and study of the cissoid, see Introduction pp. 24-25). Nevertheless, since the term is now universally employed for Diocles' curve, and since it is convenient to have a name to refer to it by, from now on I use "cissoid" to designate the curve described here.

220 P serves here as an *arbitrary* point.

221–30 Prop. 13. Use of the curve ("cissoid") to solve the problem of constructing a line, the cube on which shall be in a given proportion to the cube on a given line.

221 This is a more general formulation of the problem of doubling the cube.

223 <u>in the way we described</u> See Prop. 12.

226 See Props. 11 and 12.

<u>when four lines... the cube on the second</u> For a proof see note on 205.

227 <u>DM : MK = $DM^3 : LM^3$</u> For $\frac{DM}{ML} = \frac{ML}{MZ} = \frac{MZ}{MK}$ (by the property of the "cissoid"). Therefore ML and MZ are mean proportionals between DM and MK, therefore DM : MK = $DM^3 : LM^3$ (226).

229 I. e., we construct a rectangle of given side (S) and given area (A^2). This is possible by Euclid I 44. Then $A^2 = S.X$, and the other side of the rectangle is X (throughout this section, where I have put "X", the Arabic has "that other line").

231–34 Prop. 14. An auxiliary construction.

I regard sections 232-35 (with the accompanying Figs. 14 and 15A) as a spurious addition in the Arabic, intended to elucidate an incorrect drawing of Fig. 15B. What Diocles did (in Prop. 15, excluding 235) was to provide an alternative solution of Prop. 13, which has the advantage that instead of constructing a new circle for every given line (222), one constructs once for all a figure which will solve the problem for any given line (even if the line exceeds the side of the figure, 240). Thus 231 is clearly in place, and should be followed immediately by 236.

235–43 Prop. 15. Alternative solution of the problem of constructing a line, the cube on which shall be in a given proportion to the cube on a given line, by means of the "cissoid".

Fig. 15B should be used, and 235 neglected (see preceding note).

236 <u>the line previously mentioned</u> The "cissoid".

238 <u>wal-nujiz</u> This is the correct form of the jussive of the IVth form of "jāza", "let us make pass" (cf. nujīzu, 237). The ms. appears to have "wal-nujāza" or something similar. This could normally be interpreted only as some form of

the passive, which is clearly impossible here, where we have an accusative ("kaṭṭan") governed by the verb. However, there may well be some connection with the aberrant forms of "jāza" mentioned in the note on 45, and possibly all five occurrences (45, 113, 131, 163, 238) have the same explanation, which I am unable to provide.

239 See 227.

240 Diocles here uses what is somewhat misleadingly called "Archimedes' Axiom", in the form in which it is found in Euclid X 1, porism, namely that if any magnitude be continually halved, one will eventually obtain a magnitude less than any assigned magnitude. This (and several equivalent axioms) were used in Greek mathematics in the process of passage to the limit (often erroneously called "exhaustion" in modern times) by which ancient mathematicians solved problems involving integration. Such axioms go back at least to Eudoxus (first half of fourth century). For a discussion of this and other forms see Heath, *Euclid* III pp. 15-16 and Heath, *Archimedes*, Introduction pp. xlvii-xlviii.

242 I. e., if the ratio of E : Z (= AH : HK) is such that, although AH is less then AB, HK falls below BNS (as in Fig. VII), the procedure still works. The only difference is that the order of lines HN and MR is interchanged.

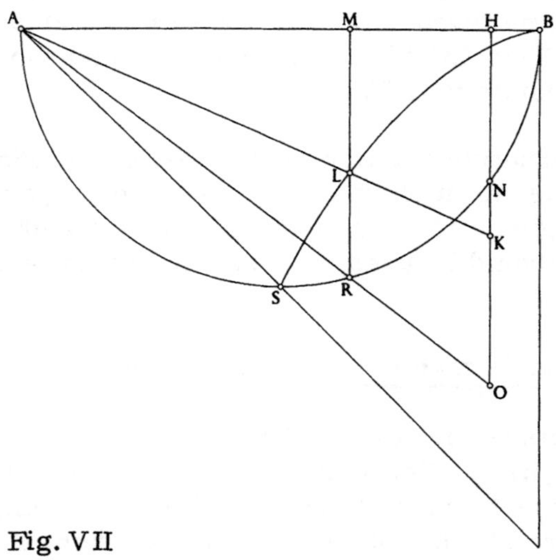

Fig. VII

243 The text is very dubious here, but the sense is clear: Diocles means "any two similar three-dimensional bodies".

244-51 Prop. 16. To find two mean proportionals between two given straight lines.

This differs from the previous propositions (13 and 15) in the following way. There the problem was: given a line x and a ratio p : q, construct a line y such that $x^3 : y^3 = p : q$. It was solved by constructing line z so that x : z = p : q, then finding lines a : d = x : z, where a and d are related to the "cissoid". Then (from the "cissoid") one has two intermediate lines b and c, so that $a : d = a^3 : b^3$, hence $p : q = x : z = a : d = a^3 : b^3$, so $y = \frac{x.b.}{a}$ Here the problem is directly: given a and d, find two mean proportionals b and c. This cannot be done immediately from the "cissoid" as constructed by Diocles, for that would require, in Fig. 15B, finding the point M such that AM = a and ML = d, which is not possible by Euclidean geometry. Hence the auxiliary construction of Prop. 16 is necessary (a similar idea is expressed by Eutocius, p. 70, 2-4, and elucidated by Heiberg, p. 71 n. 1).

244 al-arbaʿa al-k̲aṭṭ This is grammatically impossible. The simplest solution, which I have adopted, is to suppress "al-k̲aṭṭ" as a scribal gloss. The expression then becomes identical to that found immediately below. However, one could also emend "al-k̲aṭṭ" to "al-k̲uṭūṭ": see Wright II p.244B § 107(d) for this (rather rare) construction.

245 See Props. 13 and 15

252 tammat I presume that the feminine (instead of the masculine agreeing with "kitāb") is used because it is immediately preceded by the plural "al-marāyā".

243 [illegible] the text is very [illegible] [illegible] Dionysius [illegible] [illegible] [illegible] two [illegible] [illegible] bodies [illegible]

244-1 [illegible] To find [illegible] proportionals [illegible] given [illegible]

This [illegible] from the [illegible] [illegible] [illegible] [illegible] [illegible] [illegible] [illegible] [illegible] [illegible] [illegible] [illegible] rotated [illegible] [illegible] [illegible] [illegible] [illegible] lines [illegible] and [illegible] [illegible] hence [illegible]

[illegible] the [illegible]

[illegible] [illegible] [illegible] [illegible] [illegible] [illegible] [illegible] [illegible] [illegible] [illegible] which [illegible] [illegible] [illegible] [illegible] [illegible] [illegible] [illegible] [illegible] [illegible]

[illegible] [illegible] [illegible] [illegible] [illegible] [illegible] [illegible] [illegible]

[illegible] [illegible] [illegible] [illegible] [illegible] [illegible] [illegible]

[illegible] [illegible] [illegible] [illegible] [illegible] [illegible]

Appendix A

Text and translation of Eutocius' excerpts from On Burning Mirrors

Except for his version of Prop. 7 (beginning of (i) below), Eutocius does not quote Diocles verbatim, but reformulates the proofs to suit the pedantic norms of his own time (cf. Introduction p. 18). Thus he gives both "analysis" and "synthesis" for Props. 8 and 10, whereas Diocles gives only analysis for Prop. 8 and only synthesis for Prop. 10. Perhaps the most radical transformation is Eutocius' elimination of all mention of the focus-directrix construction of the parabola in Prop. 10.

The Greek text is based on that of Heiberg (*Archimedis Opera Omnia* III), but I have checked Heiberg's readings against the ms. Florence, Laur. Plut. XXVIII 4, and have occasionally adopted a different reading or emendation. The sigla in the critical apparatus have the following meanings:

A The lost manuscript of Giorgio Valla (the readings are deduced from the consensus of all or most of the mss. D, E, G and H, which are derived from A).
B The Greek text used by William of Moerbeke (deduced from the autograph of his Latin translation, Vatican, Ottobonianus lat. 1850). B^2 refers to a correction or second hand in Moerbeke's autograph, B^1 to the original reading.
Bas. The *editio princeps* of Archimedes, Basel 1544.
D Florence, Laur. Plut. XXVIII 4.
E Venice, Marc. gr. 305.
G Parisinus gr. 2360.
H Parisinus gr. 2361.

The figures in this appendix are copied (with slight emendations of the lettering) from figures in the medieval manuscripts, to provide some comparison with the reconstructed figures in Diocles' text.

ὡς Διοκλῆς ἐν τῷ περὶ πυρίων

γράφει δὲ καὶ ὁ Διοκλῆς ἐν τῷ περὶ πυρίων προλέγων τάδε· ἐν τῷ περὶ σφαίρας καὶ κυλίνδρου Ἀρχιμήδης ἀπέδειξεν ὅτι πᾶν τμῆμα σφαίρας ἴσον ἐστὶν κώνῳ τῷ βάσιν μὲν ἔχοντι τὴν αὐτὴν τῷ τμήματι, ὕψος δὲ εὐθεῖάν τινα λόγον ἔχουσαν πρὸς τὴν ἀπὸ τῆς τοῦ τμήματος κορυφῆς ἐπὶ τὴν βάσιν κάθετον, ὃν ἔχει συναμφότερος ἥ τε ἐκ τοῦ κέντρου τῆς σφαίρας καὶ ἡ τοῦ ἐναλλὰξ τμήματος κάθετος πρὸς τὴν τοῦ ἐναλλὰξ τμήματος κάθετον. οἷον, ἐὰν ᾖ σφαῖρα ἡ αβγ καὶ τμηθῇ ἐπιπέδῳ τινὶ τῷ περὶ διάμετρον τὴν γδ κύκλῳ, καὶ διαμέτρου οὔσης τῆς αβ, κέντρου δὲ τοῦ ε, ποιήσωμεν, ὡς συναμφότερον τὴν εαζ πρὸς ζα, οὕτως τὴν ηζ πρὸς ζβ, ἔτι τε ὡς συναμφότερον τὴν εβ, βζ πρὸς ζβ, οὕτως τὴν θζ πρὸς ζα, ἀποδέδεικται ὅτι τὸ μὲν γβδ τμῆμα τῆς σφαίρας ἴσον ἐστὶ τῷ κώνῳ, οὗ βάσις μέν ἐστιν ὁ περὶ διάμετρον τὴν γδ κύκλος, ὕψος δὲ ἡ ζη, τὸ δὲ γαδ τμῆμα ἴσον ἐστὶ τῷ κώνῳ, οὗ βάσις μέν ἐστιν ἡ αὐτή, ὕψος δὲ ἡ θζ. προταθέντος οὖν αὐτῷ τοῦ τὴν δοθεῖσαν σφαῖραν ἐπιπέδῳ τεμεῖν, ὥστε τὰ τμήματα τῆς σφαίρας πρὸς ἄλληλα λόγον ἔχειν τὸν δοθέντα, κατασκευάσας τὰ εἰρημένα φησί· λόγος ἄρα δοθεὶς καὶ τοῦ κώνου, οὗ βάσις ἐστὶν ὁ περὶ διάμετρον τὴν γδ κύκλος, ὕψος δὲ ἡ ζθ, πρὸς τὸν κῶνον, οὗ βάσις μέν ἐστιν ἡ αὐτή, ὕψος δε ἡ ζη· ⟨οὗτος δ' ἐστὶ λόγος τῆς θζ πρὸς ζη·⟩ καὶ γὰρ καὶ τοῦτο ἀπε-

---

5 τὴν αὐτὴν GB^2, κωνω την A 6 κορυφῆς B, κορυφην A

12 εαζ: εζ ζα codd. 13 ζα BG, ζδ A

23 οὗτος ... ζη· addidi, collato textu Dioclis Arabico

(i) Heiberg III p. 160, 2 to p. 174, 4 (corresponds to Diocles' Props. 7 and 8).

As Diocles in "On Burning Mirrors".

Diocles also, in his book "On Burning Mirrors", writes [on this problem]. His introduction is as follows.

In his book "On the Sphere and Cylinder" Archimedes proved that every segment of a sphere is equal to the cone whose base is the same as the segment, and whose height is a straight line whose ratio to the perpendicular from the vertex of the segment to its base equals the ratio of the sum of the radius of the sphere and the perpendicular of the other segment to the perpendicular of the other segment. For example, if ABΓ is a sphere, and it is cut by a plane, [namely] the circle on diameter ΓΔ, and AB is the diameter and E the center [of the sphere], and we set

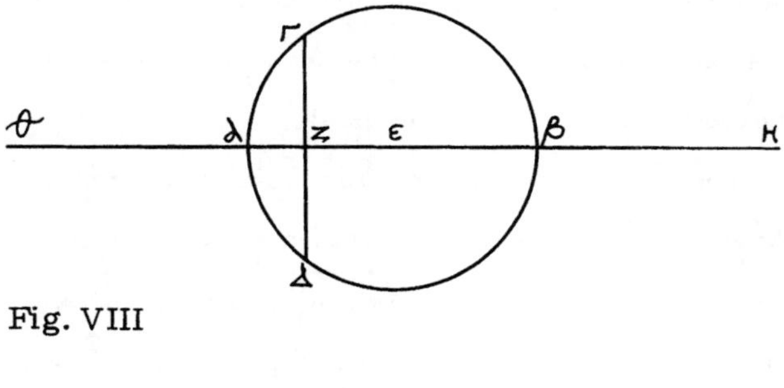

Fig. VIII

HZ : ZB = (EA + AZ) : ZA,
and also θZ : ZA = (EB + BZ) : ZB,

[then] it has been proven that segment ΓBΔ of the sphere is equal to the cone whose base is the circle on diameter ΓΔ and whose height is ZH, while segment ΓAΔ is equal to the cone whose base is the same and whose height is θZ.

So, having been set the problem of cutting the given sphere with a plane so that the segments of the sphere bear a given ratio, he [Archimedes] says, after making the above construction: "So the ratio of the cone whose base is the circle on diameter ΓΔ and whose height is Zθ to the cone whose base is the same and whose height is ZH is given. And this is the ratio of θZ to ZH. For it has also been proven that cones on equal bases have to one another the ratio of

δείχθη, ⟨ὅτι⟩ οἱ κῶνοι οἱ ἐπ' ἴσων βάσεων ὄντες πρὸς ἀλλήλους εἰσὶν ὡς τὰ ὕψη· λόγος ἄρα τῆς $\overline{\theta\zeta}$ πρὸς $\overline{\zeta\eta}$ δοθείς. καὶ ἐπεί ἐστιν, ὡς ἡ $\overline{\theta\zeta}$ πρὸς $\overline{\zeta\alpha}$, οὕτως συναμφότερος ἡ $\overline{\varepsilon\beta\zeta}$ πρὸς τὴν $\overline{\zeta\beta}$, διελόντι, ὡς ἡ $\overline{\theta\alpha}$ πρὸς $\overline{\alpha\zeta}$, οὕτως ἡ $\overline{\varepsilon\beta}$ πρὸς $\overline{\zeta\beta}$. διὰ τὰ αὐτὰ δὴ καί, ὡς ἡ $\overline{\eta\beta}$ πρὸς $\overline{\zeta\beta}$, οὕτως ἡ αὐτὴ εὐθεῖα πρὸς τὴν $\overline{\zeta\alpha}$. γέγονεν οὖν πρόβλημα τοιοῦτον· θέσει οὔσης εὐθείας τῆς $\overline{\alpha\beta}$ καὶ δύο δοθέντων σημείων τῶν $\overline{\alpha}$, $\overline{\beta}$, καὶ δοθείσης τῆς $\overline{\varepsilon\beta}$, τεμεῖν τὴν $\overline{\alpha\beta}$ κατὰ τὸ $\overline{\zeta}$ καὶ προσθεῖναι τὰς $\overline{\theta\alpha}$, $\overline{\beta\eta}$, ὥστε λόγον εἶναι τῆς $\overline{\theta\zeta}$ πρὸς $\overline{\zeta\eta}$ δοθέντα, ἔτι τε εἶναι, ὡς μὲν τὴν $\overline{\theta\alpha}$ πρὸς $\overline{\alpha\zeta}$, οὕτως τὴν δοθεῖσαν εὐθεῖαν πρὸς τὴν $\overline{\zeta\beta}$, ὡς δὲ τὴν $\overline{\eta\beta}$ πρὸς $\overline{\beta\zeta}$, οὕτως τὴν αὐτὴν δοθεῖσαν εὐθεῖαν πρὸς $\overline{\zeta\alpha}$. τοῦτο δὲ ἑξῆς δέδεικται· ὁ γὰρ Ἀρχιμήδης μακρότερον αὐτὸ δείξας καὶ οὕτως εἰς πρόβλημα ἕτερον ἀπάγει ὃ οὐκ ἀποδείκνυσιν ἐν τῷ περὶ σφαίρας καὶ κυλίνδρου.

θέσει δεδομένης εὐθείας τῆς $\overline{\alpha\beta}$ καὶ δύο δοθέντων σημείων τῶν $\overline{\alpha}$, $\overline{\beta}$, καὶ λόγου τοῦ ὃν ἔχει ἡ $\overline{\gamma}$ πρὸς τὴν $\overline{\delta}$, τεμεῖν τὴν $\overline{\alpha\beta}$ κατὰ τὸ $\overline{\varepsilon}$ καὶ προσθεῖναι τὰς $\overline{\zeta\alpha}$, $\overline{\eta\beta}$, ὥστε εἶναι, ὡς τὴν $\overline{\gamma}$ πρὸς τὴν $\overline{\delta}$, οὕτως τὴν $\overline{\zeta\varepsilon}$ πρὸς τὴν $\overline{\varepsilon\eta}$, ἔτι τε εἶναι, ὡς τὴν $\overline{\zeta\alpha}$ πρὸς $\overline{\alpha\varepsilon}$, οὕτως δοθεῖσάν τινα εὐθεῖαν πρὸς τὴν $\overline{\beta\varepsilon}$, ὡς δὲ τὴν $\overline{\eta\beta}$ πρὸς $\overline{\beta\varepsilon}$, οὕτως τὴν αὐτὴν δοθεῖσαν εὐθεῖαν πρὸς τὴν $\overline{\varepsilon\alpha}$.

γεγονέτω, καὶ τῇ $\overline{\alpha\beta}$ πρὸς ὀρθὰς ἤχθωσαν αἱ $\overline{\theta\alpha\kappa}$, $\overline{\lambda\beta\mu}$, καὶ τῇ δοθείσῃ εὐθείᾳ ἴση κείσθω ἑκατέρα τῶν $\overline{\alpha\kappa}$, $\overline{\beta\mu}$. ἐπιζευχθεῖσαι αἱ $\overline{\kappa\varepsilon}$, $\overline{\mu\varepsilon}$ ἐκβεβλήσθωσαν ἐπὶ τὰ $\overline{\lambda}$, $\overline{\theta}$, ἐπεζεύχθω

---

24 ὅτι B ("quod coni"), om. A. Cf. textum Dioclis Arabicum

27 $\overline{\zeta\beta}^{1}$ B^{2}, $\overline{\zeta\varepsilon}$ A

45 $\overline{\theta\alpha\kappa}$ B^{2}, $\overline{\theta\alpha\zeta}$ AB^{1}

47 ἐκβεβλήσθωσαν G, B ("educantur"), εκβληθεισαν A

their heights. So the ratio of θZ to ZH is given". And since

θZ : ZA = (EB + BZ) : BZ,

dirimendo,

θA : AZ = EB : BZ.

Similarly HB : BZ = the same straight line [i. e. EB] : ZA. So the problem has turned out as follows:

If line AB be given in position, and the two points A, B be given, and line EB be given [in magnitude], to cut AB at Z and add [to AB] θA and BH so that the ratio of θZ to ZH be given, and furthermore that the ratio of θA to AZ be equal to the ratio of the given line to ZB, and that the ratio of HB to BZ be equal to the ratio of the same given line to ZA. This has been solved in what follows; for Archimedes, having expounded it by a longer method, even so reduces it to another problem which he does not solve in "On the Sphere and Cylinder".

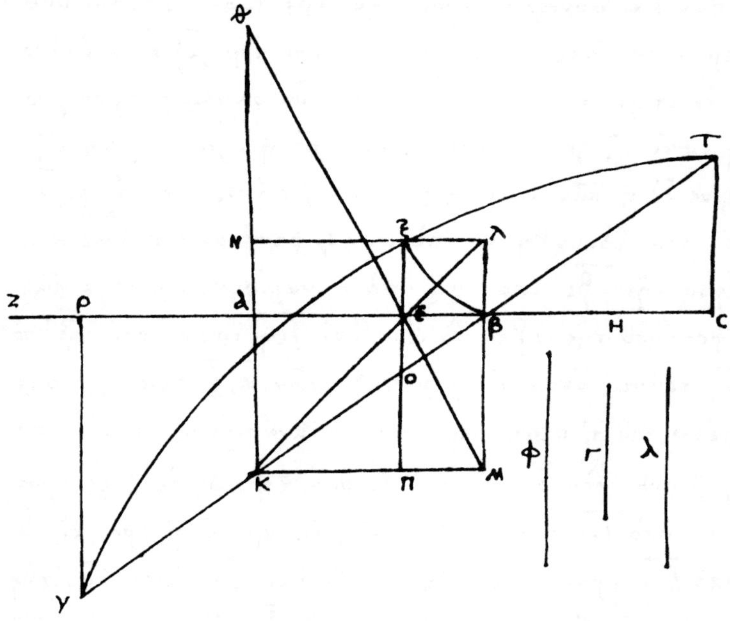

Fig. IX

When AB is given in position, and two points, A, B, are given, and the ratio of Γ to Δ [is given], to cut AB at E and add [to AB] ZA and HB, so that

ZE : EH = Γ : Δ,

and also some given straight line : BE = ZA : AE,
and the same straight line : EA = HB : BE.

Suppose it done, and let θAK, ΛBM be drawn at right angles to AB. Let both AK and BM be equal to the given straight line. Let KE,

δὲ καὶ ἡ κμ, καὶ διὰ τοῦ λ παράλληλος ἤχθω τῇ αβ ἡ λν, διὰ δὲ τοῦ ε τῇ νκ ἡ ξεοπ. ἐπεὶ οὖν ἐστιν, ὡς ἡ ζα πρὸς αε, οὕτως ἡ μβ πρὸς βε· ὑπόκειται γάρ· ὡς δὲ ἡ μβ πρὸς βε, οὕτως ἡ θα πρὸς αε διὰ τὴν ὁμοιότητα τῶν τριγώνων, ὡς ἄρα ἡ ζα πρὸς αε, οὕτως ἡ θα πρὸς αε· ἴση ἄρα ἡ ζα τῇ θα. διὰ τὰ αὐτὰ δὴ καὶ ἡ βη τῇ βλ. καὶ ἐπεί ἐστιν, ὡς συναμφότερος ἡ θαε πρὸς συναμφότερον τὴν μβε, οὕτως συναμφότερος ἡ καε πρὸς συναμφότερον τὴν λβε· ἑκάτερος γὰρ τῶν λόγων ὁ αὐτός ἐστι τῷ τῆς αε πρὸς εβ· τὸ ἄρα ὑπὸ συναμφοτέρου τῆς θαε καὶ συναμφοτέρου τῆς λβε ἴσον ἐστὶ τῷ ὑπὸ συναμφοτέρου τῆς καε καὶ συναμφοτέρου τῆς μβε. κείσθω τῇ κα ἴση ἑκατέρα τῶν αρ, βσ. ἐπεὶ οὖν συναμφότερος μὲν ἡ θαε ἴση ἐστὶ τῇ ζε, συναμφότερος δὲ ἡ λβε ἴση τῇ εη, συναμφότερος δὲ ἡ καε ἴση τῇ ρε, συναμφότερος δὲ ἡ μβε ἴση τῇ σε, καὶ ἐδείχθη τὸ ὑπὸ συναμφοτέρου τῆς θαε καὶ συναμφοτέρου τῆς λβε ἴσον τῷ ὑπὸ συναμφοτέρου τῆς καε καὶ συναμφοτέρου τῆς μβε, τὸ ἄρα ὑπὸ ζεη ἴσον ἐστὶ τῷ ὑπὸ ρεσ. διὰ δὴ τοῦτο, ὅταν τὸ ρ μεταξὺ τῶν α, ζ πίπτῃ, τότε τὸ σ ἐξωτέρω τοῦ η πεσεῖται, καὶ τὸ ἀνάπαλιν. ἐπεὶ οὖν ἐστιν, ὡς ἡ γ πρὸς τὴν δ, οὕτως ἡ ζε πρὸς εη, ὡς δὲ ἡ ζε πρὸς εη, οὕτως τὸ ὑπὸ ζεη πρὸς τὸ ἀπὸ εη, ὡς ἄρα ἡ γ πρὸς τὴν δ, οὕτως τὸ ὑπὸ ζεη πρὸς τὸ ἀπὸ εη. τὸ δὲ ὑπὸ ζεη ἴσον ἐδείχθη τῷ ὑπὸ ρεσ· ἔστιν ἄρα, ὡς ἡ γ πρὸς τὴν δ, οὕτως τὸ ὑπὸ ρεσ πρὸς τὸ ἀπὸ εη. κείσθω τῇ βε ἴση ἡ εο, καὶ ἐπιζευχθεῖσα ἡ

---

52 τῇ B^2, προς AB^1

57 θαε B, θλε A; λβε B^2, αβε A

71 εο Bas., εθ A, βο B

ME be joined and produced to Λ, Θ, and let KM also be joined, and let ΛN be drawn through Λ parallel to AB, and ΞΕΟΠ be drawn through E parallel to NK. Then since

$MB : BE = ZA : AE$ (by hypothesis)

and

$\Theta A : AE = MB : BE$ (from the similarity of the triangles),

therefore

$\Theta A : AE = ZA : AE.$

Therefore

$ZA = \Theta A.$

By the same reasoning

$BH = B\Lambda$ also.

And since

$(KA + AE) : (\Lambda B + BE) = (\Theta A + AE) : (MB + BE)$

(for each of the ratios equals AE : EB), therefore

$(\Theta A + AE) . (\Lambda B + BE) = (KA + AE) . (MB + BE).$

Let both AP and BΣ be equal to KA. Then since

$\Theta A + AE = ZE$ and $\Lambda B + BE = EH$

and

$KA + \Lambda E = PE$ and $MB + BE = \Sigma E$,

and it has been shown that

$(\Theta A + AE) . (\Lambda B + BE) = (KA + AE) . (MB + BE),$

therefore

$ZE . EH = PE . E\Sigma.$

For this reason, when P falls between A and Z, then Σ will fall outside H, and vice versa. Then since

$ZE : EH = \Gamma : \Delta$ and

$(ZE . EH) : (EH)^2 = ZE : EH,$

therefore

$(ZE . EH) : (EH)^2 = \Gamma : \Delta.$

But it was shown that

$ZE . EH = PE . E\Sigma.$

Therefore

$(PE . E\Sigma) : (EH)^2 = \Gamma : \Delta.$

Let EO be equal to BE, and let BO be joined and produced on both sides, and let ΣT, PY be drawn at right angles [to AB] and meet

$\overline{βο}$ ἐκβεβλήσθω ἐφ' ἑκάτερα, καὶ ἀπὸ τῶν $\overline{σ}$, $\overline{ρ}$ πρὸς ὀρθὰς ἀχθεῖσαι αἱ $\overline{στ}$, $\overline{ρυ}$ συμβαλλέτωσαν αὐτῇ κατὰ τὰ $\overline{τ}$, $\overline{υ}$. ἐπεὶ οὖν διὰ δεδομένου τοῦ $\overline{β}$ πρὸς θέσει δεδομένην τὴν $\overline{αβ}$ ἦκται ἡ $\overline{τυ}$ δεδομένην ποιοῦσα γωνίαν τὴν ὑπὸ $\overline{εβο}$ ἡμισεῖαν ὀρθῆς, δέδοται ἡ $\overline{τυ}$ τῇ θέσει. καὶ ἀπὸ δεδομένων τῶν $\overline{σ}$, $\overline{ρ}$ θέσει ἠγμέναι αἱ $\overline{στ}$, $\overline{ρυ}$ τέμνουσιν αὐτὴν κατὰ τὰ $\overline{τ}$, $\overline{υ}$. δοθέντα ἄρα ἐστὶ τὰ $\overline{τ}$, $\overline{υ}$· δοθεῖσα ἄρα ἐστὶν ἡ $\overline{τυ}$ τῇ θέσει καὶ τῷ μεγέθει. καὶ ἐπεὶ διὰ τὴν τῶν $\overline{εοβ}$, $\overline{στβ}$ τριγώνων ὁμοιότητά ἐστιν, ὡς ἡ $\overline{τβ}$ πρὸς $\overline{βο}$, οὕτως ἡ $\overline{σβ}$ πρὸς $\overline{βε}$, καὶ συνθέντι ἐστίν, ὡς ἡ $\overline{το}$ πρὸς $\overline{οβ}$, οὕτως ἡ $\overline{σε}$ πρὸς $\overline{εβ}$. ἀλλ' ὡς ἡ $\overline{βο}$ πρὸς $\overline{ου}$, οὕτως ἡ $\overline{βε}$ πρὸς $\overline{ερ}$· καὶ δι' ἴσου ἄρα, ὡς ἡ $\overline{το}$ πρὸς $\overline{ου}$, οὕτως ἡ $\overline{σε}$ πρὸς $\overline{ερ}$. ἀλλ' ὡς ἡ $\overline{το}$ πρὸς $\overline{ου}$, οὕτως τὸ ὑπὸ $\overline{του}$ πρὸς τὸ ἀπὸ $\overline{ου}$, ὡς δὲ ἡ $\overline{σε}$ πρὸς $\overline{ερ}$, οὕτως τὸ ὑπὸ $\overline{σερ}$ πρὸς τὸ ἀπὸ $\overline{ερ}$· καὶ ὡς ἄρα τὸ ὑπὸ $\overline{του}$ πρὸς τὸ ἀπὸ $\overline{ου}$, οὕτως τὸ ὑπὸ $\overline{σερ}$ πρὸς τὸ ἀπὸ $\overline{ερ}$· καὶ ἐναλλάξ, ὡς τὸ ὑπὸ $\overline{του}$ πρὸς τὸ ὑπὸ $\overline{σερ}$, οὕτως τὸ ἀπὸ $\overline{ου}$ πρὸς τὸ ἀπὸ $\overline{ερ}$. τὸ δὲ ἀπὸ $\overline{ου}$ τοῦ ἀπὸ $\overline{ερ}$ διπλάσιον, ἐπειδὴ καὶ τὸ ἀπὸ $\overline{οβ}$ τοῦ ἀπὸ $\overline{βε}$· καὶ τὸ ὑπὸ $\overline{του}$ ἄρα τοῦ ὑπὸ $\overline{σερ}$ ἐστι διπλάσιον. τὸ δὲ ὑπὸ $\overline{σερ}$ πρὸς τὸ ἀπὸ $\overline{εη}$ ἐδείχθη λόγον ἔχειν, ὃν ἔχει ἡ $\overline{γ}$ πρὸς τὴν $\overline{δ}$· καὶ τὸ ὑπὸ $\overline{του}$ ἄρα πρὸς τὸ ἀπὸ $\overline{εη}$ λόγον ἔχει, ὃν ἡ διπλασία τῆς $\overline{γ}$ πρὸς τὴν $\overline{δ}$. τὸ δὲ ἀπὸ $\overline{εη}$ ἴσον ἐστὶ τῷ

---

75 ποιοῦσα Heib., ποιουσαν codd.

77 $\overline{στ}$, $\overline{ρυ}$ Bas., $\overline{σρ}$ $\overline{τυ}$ A

79 $\overline{στβ}$ B, $\overline{στυ}$ A

80 $\overline{σβ}$ πρὸς $\overline{βε}$ B^2, εβ προς $\overline{βσ}$ A   83 $\overline{ου}^1$ B^2, $\overline{ουσ}$ A

85-86 καὶ ... $\overline{ερ}$ bis A (in rep. 86 απο pro ὑπὸ)

90 $\overline{γ}$ B, $\overline{ηγ}$ A

it [BO produced] at T, Y. Then since TY has been drawn through a given point, B, to a straight line given in position, AB, making a given angle [with AB], EBO, (half a right angle), TY is given in position. And ΣT, PY, drawn from points given in position, Σ , P, cut it at T, Y. Therefore T and Y are given. Therefore TY is given in position and magnitude. And since it follows from the similarity of the triangles EOB, ΣTB that

$$\Sigma B : BE = TB : BO,$$

componendo

$$\Sigma E : EB = TO : OB.$$

But

$$BE : EP = BO : OY,$$

so *ex aequo*

$$\Sigma E : EP = TO : OY.$$

But $(TO . OY) : (OY)^2 = TO : OY$ and $(\Sigma E . EP) : (EP)^2 = \Sigma E : EP$, therefore $(\Sigma E . EP) : (EP)^2 = (TO . OY) : (OY)^2$, and *permutando*, $(OY)^2 : (EP)^2 = (TO . OY) : (\Sigma E . EP)$.

But $(OY)^2 = 2(EP)^2$, since $(OB)^2 = 2(BE)^2$, therefore $(TO . OY) = 2(\Sigma E . EP)$.

But it was shown that

$$(\Sigma E . EP) : (EH)^2 = \Gamma : \Delta.$$

ἀπὸ $\overline{\xi o}$· ἑκατέρα γὰρ τῶν $\overline{\epsilon\eta}$, $\overline{\xi o}$ ἴση ἐστὶ συναμφοτέρῳ τῇ $\overline{\lambda\beta\epsilon}$· τὸ ἄρα ὑπὸ $\overline{\tau o\upsilon}$ πρὸς τὸ ἀπὸ $\overline{\xi o}$ λόγον ἔχει, ὃν ἡ διπλασία τῆς $\overline{\gamma}$ πρὸς τὴν $\overline{\delta}$. καὶ δέδοται ὁ τῆς διπλασίας τῆς $\overline{\gamma}$ πρὸς τὴν $\overline{\delta}$ λόγος· δέδοται ἄρα καὶ ὁ τοῦ ὑπὸ $\overline{\tau o\upsilon}$ πρὸς τὸ ἀπὸ $\overline{\xi o}$ λόγος. ἐὰν ἄρα ποιήσωμεν, ὡς τὴν $\overline{\delta}$ πρὸς τὴν διπλασίαν τῆς $\overline{\gamma}$, οὕτως τὴν $\overline{\tau\upsilon}$ πρὸς ἄλλην τινὰ ὡς τὴν $\overline{\phi}$, καὶ περὶ τὴν $\overline{\tau\upsilon}$ γράψωμεν ἔλλειψιν, ὥστε τὰς καταγομένας ἐν τῇ ὑπὸ $\overline{\xi o\beta}$ γωνίᾳ, τουτέστιν ἐν ἡμισείᾳ ὀρθῆς, δύνασθαι τὰ παρὰ τὴν $\overline{\phi}$ ἐλλείποντα ὁμοίῳ τῷ ὑπὸ $\overline{\tau\upsilon}$, $\overline{\phi}$, ἥξει διὰ τοῦ $\overline{\xi}$ διὰ τὴν ἀντιστροφὴν τοῦ εἰκοστοῦ θεωρήματος τοῦ πρώτου βιβλίου τῶν Ἀπολλωνίου κωνικῶν στοιχείων. γεγράφθω καὶ ἔστω ὡς ἡ $\overline{\upsilon\xi\tau}$· τὸ ἄρα $\overline{\xi}$ σημεῖον ἅπτεται θέσει δεδομένης ἐλλείψεως. καὶ ἐπεὶ διαγώνιός ἐστιν ἡ $\overline{\lambda\kappa}$ τοῦ $\overline{\nu\mu}$ παραλληλογράμμου, ἴσον ἐστὶ τὸ ὑπὸ $\overline{\nu\xi\pi}$ τῷ ὑπὸ $\overline{\alpha\beta\mu}$. ἐὰν ἄρα διὰ τοῦ $\overline{\beta}$ περὶ ἀσυμπτώτους τὰς $\overline{\theta\kappa\mu}$ γράψωμεν ὑπερβολήν, ἥξει διὰ τοῦ $\overline{\xi}$, καὶ ἔσται θέσει δεδομένη διὰ τὸ καὶ τὸ $\overline{\beta}$ σημεῖον τῇ θέσει δεδόσθαι καὶ ἑκατέραν τῶν $\overline{\alpha\beta}$, $\overline{\beta\mu}$ καὶ διὰ τοῦτο τὰς $\overline{\theta\kappa\mu}$ ἀσυμπτώτους. γεγράφθω καὶ ἔστω ὡς ἡ $\overline{\xi\beta}$· τὸ ἄρα $\overline{\xi}$ σημεῖον ἅπτεται θέσει δεδομένης ὑπερβολῆς. ἥπτετο δὲ καὶ θέσει δεδομένης ἐλλείψεως· δέδοται ἄρα τὸ $\overline{\xi}$. καὶ ἀπ' αὐτοῦ κάθετος ἡ $\overline{\xi\epsilon}$· δέδοται ἄρα τὸ $\overline{\epsilon}$. καὶ ἐπεί ἐστιν, ὡς ἡ $\overline{\mu\beta}$ πρὸς $\overline{\beta\epsilon}$, οὕτως ἡ $\overline{\xi\alpha}$ πρὸς $\overline{\alpha\epsilon}$, καὶ δέδοται ἡ $\overline{\alpha\epsilon}$, δέδοται ἄρα καὶ ἡ $\overline{\alpha\xi}$.

---

93 $\overline{\xi o}$[1] BEG, $\overline{\xi\epsilon}$ A

104 $\overline{\upsilon\xi\tau}$ B^2, $\overline{\gamma\xi\tau}$ A

105 ἐλλείψεως GH, ελληψεως A

109 δεδόσθαι G, δεδοσθω A

114 $\overline{\beta\epsilon}$ B^2, $\overline{\alpha\epsilon}$ A

Therefore

$$(TO.OY) : (EH)^2 = 2\Gamma : \Delta.$$

But

$$(EH)^2 = (\Xi O)^2,$$

for both EH and ΞO are equal to (ΛB + BE).
Therefore $(TO.OY) : (\Xi O)^2 = 2\Gamma : \Delta$.
And the ratio $2\Gamma : \Delta$ is given, therefore the ratio $(TO.OY) : (\Xi O)^2$ is also given. So if we make

$$TY : \text{another line, e.g. } \Phi = \Delta : 2\Gamma,$$

and draw an ellipse about TY, so that the ordinates drawn [to TY] at the angle ΞOB, i.e. half a right angle, are equal in square to rectangles applied to Φ and falling short of it by a rectangle similar to TY.Φ, it [the ellipse] will pass through Ξ, by the converse of the 20th theorem of Book I of Apollonius' *Conic Elements*. Let it be drawn as YΞT. Then point Ξ lies on the perimeter of an ellipse given in position. And since ΛK is the diagonal of parallelogram NM,

$$N\Xi . \Xi\Pi = AB.BM.$$

So if we draw a hyperbola through B to the asymptotes ΘK, KM, it will pass through Ξ, and it [the hyperbola] will be given in position, since point B too is given in position, and so are both AB and BM, and hence the asymptotes ΘK, KM. Let it be drawn as ΞB. So point Ξ lies on the perimeter of a given hyperbola. But we found that it also lies on the perimeter of a given ellipse. Therefore Ξ is given, and ΞE is a perpendicular from it, so E is given. And since

$$ZA : AE = MB : BE$$

διὰ τὰ αὐτὰ δὴ δέδοται καὶ ἡ ηβ.

συντεθήσεται δὲ οὕτως· ὡς γὰρ ἐπὶ τῆς αὐτῆς καταγραφῆς ἔστω ἡ δοθεῖσα εὐθεῖα, ἣν δεῖ τεμεῖν, ἡ αβ, ἡ δὲ δοθεῖσα ἑτέρα ἡ ακ, ὁ δὲ δοθεὶς λόγος ὁ τῆς γ πρὸς τὴν δ. ἤχθω τῇ αβ πρὸς ὀρθὰς ἡ βμ ἴση οὖσα τῇ ακ, καὶ ἐπεζεύχθω ἡ κμ, καὶ τῇ μὲν κα ἴση κείσθω ἡ αρ καὶ ἡ βσ, ἀπὸ <δὲ> τῶν ρ, σ πρὸς ὀρθὰς ἤχθωσαν αἱ ρυ, στ, καὶ πρὸς τῷ β σημείῳ συνεστάτω ἡμισεῖα ὀρθῆς ἡ ὑπὸ αβο, καὶ ἐκβληθεῖσα ἡ βο ἐφ' ἑκάτερα τεμνέτω τὰς στ, ρυ κατὰ τὰ τ, υ, καὶ γεγονέτω, ὡς ἡ δ πρὸς τὴν διπλασίαν τῆς γ, οὕτως ἡ τυ πρὸς τὴν φ, καὶ περὶ τὴν τυ γεγράφθω ἔλλειψις, ὥστε τὰς καταγομένας ἐν ἡμισείᾳ ὀρθῆς δύνασθαι τὰ παρακείμενα παρὰ τὴν φ ἐλλείποντα ὁμοίῳ τῷ ὑπὸ τυ, φ, διὰ δὲ τοῦ β περὶ ἀσυμπτώτους τὰς ακ, κμ γεγράφθω ὑπερβολὴ ἡ βξ τέμνουσα τὴν ἔλλειψιν κατὰ τὸ ξ, καὶ ἀπὸ τοῦ ξ ἐπὶ τὴν αβ κάθετος ἤχθω ἡ ξε καὶ ἐκβεβλήσθω ἐπὶ τὸ π, διὰ δὲ τοῦ ξ τῇ αβ παράλληλος ἤχθω ἡ λξν, καὶ ἐκβεβλήσθωσαν αἱ κα, μβ ἐπὶ τὰ λ, θ, καὶ ἡ με ἐπιζευχθεῖσα ἐκβεβλήσθω καὶ συμπιπτέτω τῇ κν κατὰ τὸ θ. ἐπεὶ οὖν ὑπερβολή ἐστιν ἡ βξ, ἀσύμπτωτοι δὲ αἱ θκ, κμ, ἴσον ἐστὶ τὸ ὑπὸ νξπ τῷ ὑπὸ αβμ διὰ τὸ η' θεώρημα τοῦ δευτέρου βιβλίου τῶν Ἀπολλωνίου κωνικῶν στοιχείων, καὶ διὰ τοῦτο εὐθεῖά ἐστιν ἡ κελ. κείσθω οὖν τῇ μὲν θα ἴση ἡ αζ, τῇ δὲ λβ ἴση ἡ βη. ἐπεὶ οὖν ἐστιν, ὡς ἡ διπλασία τῆς γ πρὸς τὴν δ, οὕτως ἡ φ πρὸς τὴν τυ, ὡς δὲ ἡ φ πρὸς τὴν τυ, οὕτως τὸ ὑπὸ του πρὸς

---

118 αβ GB[2], δβ AB[1]

121 δὲ add. Heib., om. A. Cf. "et a" B[2]

128 ακ B[2], αβ A

132 κα B[2], κμ A

and AE is given, AZ too is given. By the same argument HB is given.

The synthesis will be as follows. Using the same figure, let the given line, which is to be cut, be AB, and the other given line AK, and the given ratio Γ : Δ. Let BM be drawn perpendicular to AB equal to AK. Join KM, and let AP and BΣ be equal to KA. Let PY, ΣT be drawn perpendicular from P and Σ, and let half a right angle, ABO, be constructed at point B, and let BO be produced on both sides and cut ΣT, PY at T and Y. Let

$$TY : \Phi = \Delta : 2\Gamma,$$

and let an ellipse be drawn about TY, so that the ordinates drawn [to TY] at half a right angle are equal in square to the rectangles applied to Φ and falling short of it by a rectangle similar to TY.Φ. Let a hyperbola BΞ be drawn through B to the asymptotes AK, KM, cutting the ellipse at Ξ, and let the perpendicular ΞE be drawn from Ξ to AB and produced to Π, and let ΛΞN be drawn through Ξ parallel to AB, and let KA, MB be produced to Λ, Θ. Join ME and produce it: let it meet KN at Θ. Then since BΞ is a hyperbola, and ΘK, KM its asymptotes,

$$N\Xi . \Xi\Pi = AB . BM,$$

by the eighth theorem of Book II of Apollonius' *Conic Elements*. Therefore KEΛ is a straight line. So let AZ be equal to ΘA, and BH equal to ΛB. Then since

$$\Phi : TY = 2\Gamma : \Delta$$

and $(TO . OY) : (\Xi O)^2 = \Phi : TY$[1], by the 20th theorem of Book I of

---

1 Eutocius has committed a foolish mathematical error here. The equation of the ellipse (cf. Introduction p.6) is

$$\frac{y^2}{x_1 x_2} = \frac{p}{a};$$

where p is the parameter and a the (total) diameter. In Eutocius' figure (Fig.IX) Φ corresponds to p, TY to a. So this equation should be

$$(TO.OY):(\Xi O)^2 = TY:\Phi.$$

The error is easily remedied by setting

$$TY:\Phi = 2\Gamma:\Delta$$

in the earlier definition of Φ (i.e. by inverting Eutocius' ratio). Heath *(Archimedes* p.77 n.) noted this blunder and correctly attributed it to Eutocius.

τὸ ἀπὸ $\overline{\xi o}$ διὰ τὸ κʹ θεώρημα τοῦ πρώτου βιβλίου τῶν Ἀπολλωνίου κωνικῶν στοιχείων, ὡς ἄρα ἡ διπλασία τῆς $\overline{\gamma}$ πρὸς τὴν $\overline{\delta}$, οὕτως τὸ ὑπὸ $\overline{\tau o \upsilon}$ πρὸς τὸ ἀπὸ $\overline{\xi o}$. καὶ ἐπεί ἐστιν, ὡς ἡ $\overline{\tau\beta}$ πρὸς $\overline{\beta o}$, οὕτως ἡ $\overline{\sigma\beta}$ πρὸς $\overline{\beta\epsilon}$, καὶ συνθέντι, ὡς ἡ $\overline{\tau o}$ πρὸς $\overline{o\beta}$, οὕτως ἡ $\overline{\sigma\epsilon}$ πρὸς $\overline{\epsilon\beta}$. ἀλλ' ὡς ἡ $\overline{\beta o}$ πρὸς $\overline{o\upsilon}$, οὕτως ἡ $\overline{\beta\epsilon}$ πρὸς $\overline{\epsilon\rho}$· καὶ δι' ἴσου ἄρα, ὡς ἡ $\overline{\tau o}$ πρὸς $\overline{o\upsilon}$, οὕτως ἡ $\overline{\sigma\epsilon}$ πρὸς $\overline{\epsilon\rho}$. καὶ ὡς ἄρα τὸ ὑπὸ $\overline{\tau o\upsilon}$ πρὸς τὸ ἀπὸ $\overline{o\upsilon}$, οὕτως τὸ ὑπὸ $\overline{\sigma\epsilon\rho}$ πρὸς τὸ ἀπὸ $\overline{\epsilon\rho}$· ἐναλλάξ, ὡς τὸ ὑπὸ $\overline{\tau o\upsilon}$ πρὸς τὸ ὑπὸ $\overline{\sigma\epsilon\rho}$, οὕτως τὸ ἀπὸ $\overline{o\upsilon}$ πρὸς τὸ ἀπὸ $\overline{\epsilon\rho}$. ἀλλὰ τὸ ἀπὸ $\overline{o\upsilon}$ τοῦ ἀπὸ $\overline{\epsilon\rho}$ διπλάσιον διὰ τὸ καὶ τὸ ἀπὸ $\overline{\beta o}$ τοῦ ἀπὸ $\overline{\beta\epsilon}$· ἴση γάρ ἐστιν ἡ $\overline{\beta\epsilon}$ τῇ $\overline{\epsilon o}$ ἡμισείας ὀρθῆς οὔσης ἑκατέρας τῶν πρὸς τοῖς $\overline{\beta}$, $\overline{o}$· καὶ τὸ ὑπὸ $\overline{\tau o\upsilon}$ ἄρα διπλάσιόν ἐστι τοῦ ὑπὸ $\overline{\sigma\epsilon\rho}$. ἐπεὶ οὖν ἐδείχθη, ὡς ἡ διπλασία τῆς $\overline{\gamma}$ πρὸς τὴν $\overline{\delta}$, οὕτως τὸ ὑπὸ $\overline{\tau o\upsilon}$ πρὸς τὸ ἀπὸ $\overline{\xi o}$, καὶ τῶν ἡγουμένων τὰ ἡμίση· ὡς ἄρα ἡ $\overline{\gamma}$ πρὸς τὴν $\overline{\delta}$, οὕτως τὸ ὑπὸ $\overline{\rho\epsilon\sigma}$ πρὸς τὸ ἀπὸ $\overline{\xi o}$, τουτέστι πρὸς τὸ ἀπὸ $\overline{\epsilon\eta}$· ἴση γὰρ ἡ $\overline{\xi o}$ τῇ $\overline{\epsilon\eta}$ διὰ τὸ ἑκατέραν αὐτῶν ἴσην εἶναι συναμφοτέρῳ τῇ $\overline{\lambda\beta\epsilon}$. ἐπεὶ οὖν ἐστιν, ὡς συναμφότερος ἡ $\overline{\theta\alpha\epsilon}$ πρὸς συναμφότερον τὴν $\overline{\mu\beta\epsilon}$, οὕτως συναμφότερος ἡ $\overline{\kappa\alpha\epsilon}$ πρὸς συναμφότερον τὴν $\overline{\lambda\beta\epsilon}$· ἑκάτερος γὰρ τῶν λόγων ὁ αὐτός ἐστι τῷ τῆς $\overline{\alpha\epsilon}$ πρὸς $\overline{\epsilon\beta}$· τὸ ἄρα ὑπὸ συναμφοτέρου τῆς $\overline{\theta\alpha\epsilon}$ καὶ συναμφοτέρου τῆς $\overline{\lambda\beta\epsilon}$ ἴσον ἐστὶ τῷ ὑπὸ συναμφοτέ-

---

146 $\overline{\sigma\epsilon\rho}$ B, $\overline{\sigma\rho\epsilon}$ A

148 $\overline{o\upsilon}^{1}$ B^{2}, $\overline{o\rho}$ A

150 οὔσης G, B^{2} ("existente"), ισης A

153 $\overline{\xi o}$ B, $\overline{\xi\theta}$ A

154–55 τουτέστι... $\overline{\xi o}$ mg. B^{2}, om. AB^{1} 156 ὡς Bas., ως η A

Apollonius' *Conic Elements*, therefore

$(TO . OY) : (\Xi O)^2 = 2\Gamma : \Delta.$

And since

$\Sigma B : BE = TB : BO,$

componendo,

$\Sigma E : EB = TO : OB.$

But

$BE : EP = BO : OY,$

so *ex aequo*

$\Sigma E : EP = TO : OY.$

Therefore

$(\Sigma E . EP) : (EP)^2 = (TO . OY) : (OY)^2.$

Permutando,

$(OY)^2 : (EP)^2 = (TO . OY) : (\Sigma E . EP).$

But $(OY)^2 = 2(EP)^2$, because $(BO)^2 = 2(BE)^2$, for BE = EO, since both the angle at B and the angle at O are half a right angle. Therefore

$TO . OY = 2(\Sigma E . EP).$

Now since it was shown that

$(TO . OY) : (\Xi O)^2 = 2\Gamma : \Delta,$

and the halves of the first members [of the ratios are] also [in the same proportion], therefore

$\Gamma : \Delta = (PE . E\Sigma) : (\Xi O)^2 = (PE . E\Sigma) : (EH)^2,$

for $\Xi O = EH$, since each is equal to $(\Lambda B + BE)$.

So since

$(KA + AE) : (\Lambda B + BE) = (\Theta A + AE) : (MB + BE),$

for each of the ratios equals AE : EB, therefore

$(\Theta A + AE) . (\Lambda B + BE) = (KA + AE) . (MB + BE).$

ρου τῆς κaε καὶ συναμφοτέρου τῆς μβε. ἀλλὰ συναμφοτέρῳ μὲν τῇ θαε ἴση ἐστὶν ἡ ζε, συναμφοτέρῳ δὲ τῇ λβε ἴση ἡ εη, συναμφοτέρῳ ⟨δὲ⟩ τῇ καε ἴση ἡ ρε, συναμφοτέρῳ δὲ τῇ μβε ἴση ἡ εσ· τὸ ἄρα ὑπὸ ζεη ἴσον ἐστὶ τῷ ὑπὸ ρεσ. ἀλλ' ὡς ἡ γ πρὸς τὴν δ, οὕτως τὸ ὑπὸ ρεσ πρὸς τὸ ἀπὸ εη. καὶ ὡς ἄρα ἡ γ πρὸς τὴν δ, οὕτως τὸ ὑπὸ ζεη πρὸς τὸ ἀπὸ εη. ἀλλ' ὡς τὸ ὑπὸ ζεη πρὸς τὸ ἀπὸ εη, οὕτως ἡ ζε πρὸς εη. καὶ ὡς ⟨ἄρα⟩ ἡ γ πρὸς τὴν δ, οὕτως ἡ ζε πρὸς εη. καὶ ἐπεί ἐστιν, ὡς ἡ μβ πρὸς βε, οὕτως ἡ θα πρὸς αε, ἴση δὲ ἡ θα τῇ ζα, ὡς ἄρα ἡ μβ πρὸς βε, οὕτως ἡ ζα πρὸς αε. διὰ τὰ αὐτὰ καί, ὡς ἡ κα πρὸς αε, οὕτως ἡ ηβ πρὸς βε. εὐθείας ἄρα δοθείσης τῆς αβ καὶ ἑτέρας τῆς ακ καὶ λόγου τοῦ τῆς γ πρὸς τὴν δ εἴληπται ἐπὶ τῆς αβ τυχὸν σημεῖον τὸ ε, καὶ προσετέθησαν εὐθεῖαι αἱ ζα, ηβ, καὶ γέγονεν ⟨ἐν⟩ τῷ δοθέντι λόγῳ ἡ ζε πρὸς εη, ἔτι τέ ἐστιν, ὡς ἡ δοθεῖσα ἡ μβ πρὸς βε, οὕτως ἡ ζα πρὸς αε, ὡς δὲ αὐτὴ ἡ δοθεῖσα ἡ κα πρὸς αε, οὕτως ἡ ηβ πρὸς βε· ὅπερ ἔδει ποιῆσαι.

Ἔστωσαν αἱ δοθεῖσαι δύο εὐθεῖαι πρὸς ὀρθὰς ἀλλήλαις αἱ αβ, βγ, καὶ γεγονέτωσαν αὐτῶν μέσαι αἱ δβ, βε, ὥστε εἶναι, ὡς τὴν γβ πρὸς βδ, οὕτως τὴν βδ πρὸς βε καὶ τὴν βε πρὸς βα, καὶ ἤχθωσαν πρὸς ὀρθὰς αἱ δζ, εζ. ἐπεὶ οὖν ἐστιν, ὡς ἡ γβ πρὸς βδ, ἡ δβ πρὸς βε, τὸ ἄρα ὑπὸ γβε, τουτέστι τὸ ὑπὸ δοθείσης καὶ

---

163 δὲ B ("ueto"), om. A

164 ὑπό² DG, απο AB

167 ἄρα B², om. AB

174 ἐν B, om. A

But

$\Theta A + AE = ZE, \Lambda B + BE = EH,$

$KA + AE = PE, MB + BE = E\Sigma.$

Therefore

$ZE . EH = PE . E\Sigma.$

But

$(PE . E\Sigma) : (EH)^2 = \Gamma : \Delta.$

Therefore

$(ZE . EH) : (EH)^2 = \Gamma : \Delta.$

But

$ZE : EH = (ZE . EH) : (EH)^2,$

therefore

$ZE : EH = \Gamma : \Delta.$

And since

$\Theta A : AE = MB : BE$

and

$\Theta A = ZA,$

therefore

$ZA : AE = MB : BE.$

By the same reasoning,

$HB : BE = KA : AE.$

Therefore, given the straight line AB and the second straight line AK, and the ratio $\Gamma : \Delta$, we have taken an arbitrary [sic!] point E on AB, and added the straight lines ZA and HB, so that ZE is in the given ratio to EH, and furthermore, the ratio of ZA to AE is as the given line MB to BE, and the ratio of HB to BE is as the given line KA to AE.

Q.E.F.

(ii) Heiberg III p. 82, 2 to 84, 7 (corresponds to Diocles' Prop. 10). On the misattribution of this solution to Menaechmus see note on Prop. 10, pp. 169-70.

Let the two given straight lines be AB, BΓ, at right angles to each other, and let ΔB, BE be [two] mean proportionals between them, so that

$\Gamma B : B\Delta = B\Delta : BE = BE : BA.$

Draw ΔZ, EZ at right angles [to BΔ, BE]. Then since

$\Delta B : BE = \Gamma B : B\Delta,$

τῆς βε, ἴσον ἐστὶ τῷ ἀπὸ τῆς βδ, τουτέστι τῆς εζ. ἐπεὶ οὖν τὸ ὑπὸ δοθείσης καὶ τῆς βε ἴσον ἐστὶ τῷ ἀπὸ εζ, τὸ ζ ἄρα ἅπτεται παραβολῆς τῆς περὶ ἄξονα τὴν βε. πάλιν, ἐπεί ἐστιν, ὡς ἡ αβ πρὸς βε, ἡ βε πρὸς βδ, τὸ ἄρα ὑπὸ αβδ, τουτέστι τὸ ὑπὸ δοθείσης καὶ τῆς βδ, ἴσον ἐστὶ τῷ ἀπὸ εβ, τουτέστι τῆς δζ. τὸ ζ ἄρα ἅπτεται παραβολῆς τῆς περὶ ἄξονα τὴν βδ. ἧπται δὲ καὶ ἑτέρας δοθείσης τῆς περὶ τὴν βε· δοθὲν ἄρα τὸ ζ. καὶ κάθετοι αἱ ζδ, ζε· δοθέντα ἄρα τὰ δ, ε.

συντεθήσεται δὲ οὕτως· ἔστωσαν αἱ δοθεῖσαι δύο εὐθεῖαι πρὸς ὀρθὰς ἀλλήλαις αἱ αβ, βγ, καὶ ἐκβεβλήσθωσαν ἐπ' ἄπειρον ἀπὸ τοῦ β, καὶ γεγράφθω περὶ ἄξονα τὴν βε παραβολή, ὥστε τὰς καταγομένας ἐπὶ τὴν βε δύνασθαι [τὰ] παρὰ τὴν βγ. πάλιν γεγράφθω περὶ ἄξονα τὴν δβ παραβολή, ὥστε τὰς καταγομένας δύνασθαι παρὰ τὴν αβ· τεμοῦσιν δὴ ἀλλήλας αἱ παραβολαί· τεμνέτωσαν κατὰ τὸ ζ, καὶ ἀπὸ τοῦ ζ κάθετοι ἤχθωσαν αἱ ζδ, ζε. ἐπεὶ οὖν ἐν παραβολῇ κατῆκται ἡ ζε, τουτέστιν ἡ δβ, τὸ ἄρα ὑπὸ γβε ἴσον ἐστὶ τῷ ἀπὸ βδ. ἔστιν ἄρα, ὡς ἡ γβ πρὸς βδ, ἡ δβ πρὸς βε. πάλιν, ἐπεὶ ἐν παραβολῇ κατῆκται ἡ ζδ, τουτέστιν ἡ εβ, τὸ ἄρα ὑπὸ δβα ἴσον ἐστὶ τῷ ἀπὸ

---

8 τὴν mg. G, της A

10 δζ B, εζ A

12 βε B, βζ A

17 τὰ seclusi

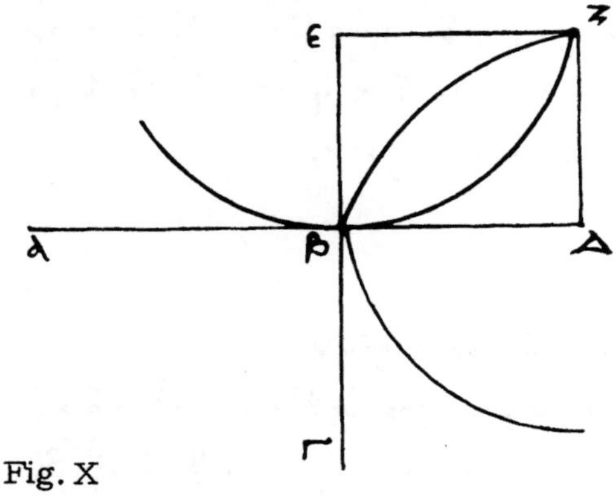

Fig. X

therefore ΓB . BE, i.e. the rectangle formed by a given line and BE, equals $B\Delta^2$, i.e. EZ^2. So, since the rectangle formed by a given line and BE equals the square on EZ, then Z lies on the circumference of the parabola with axis BE. Again, since

$$BE : B\Delta = AB : BE,$$

therefore AB . BΔ, i.e. the rectangle formed by a given line and BΔ, equals EB^2, i.e. ΔZ^2. So Z lies on the circumference of the parabola with axis BΔ. And we found that it lies on the circumference of another given [parabola] with axis BE. So Z is given, and so are the perpendiculars ZΔ and ZE. Therefore Δ and E are given.

The synthesis will be as follows. Let the two given straight lines be AB, BΓ, at right angles to each other, and let them be produced indefinitely away from B; let there be constructed on axis BE a parabola such that the ordinates to BE have as parameter BΓ. Again let there be constructed on axis ΔB a parabola such that the ordinates have AB as parameter. Now the parabolas will cut one another: let them cut at Z, and draw from Z perpendiculars ZΔ and ZE. Then since ZE (which equals ΔB) is an ordinate to a parabola, therefore

$$\Gamma B . BE = B\Delta^2.$$

Therefore

$$\Gamma B : B\Delta = \Delta B : BE.$$

Again, since ZΔ (which equals EB) is an ordinate to a parabola,

$$\Delta B . BA = EB^2.$$

εβ. ἔστιν ἄρα, ὡς ἡ δβ πρὸς βε, ἡ βε πρὸς βα. ἀλλ' ὡς ἡ δβ πρὸς βε, οὕτως ἡ γβ πρὸς βδ. καὶ ὡς ἄρα ἡ γβ πρὸς βδ, ἡ βδ πρὸς βε καὶ ἡ εβ πρὸς βα· ὅπερ ἔδει εὑρεῖν.

ὡς Διοκλῆς ἐν τῷ περὶ πυρίων

ἐν κύκλῳ ἤχθωσαν δύο διάμετροι πρὸς ὀρθὰς αἱ αβ, γδ, καὶ δύο περιφέρειαι ἴσαι ἀπειλήφθωσαν ἐφ' ἑκάτερα τοῦ β αἱ εβ, βζ, καὶ διὰ τοῦ ζ παράλληλος τῇ αβ ἤχθω ἡ ζη, καὶ ἐπεζεύχθω ἡ δε. λέγω ὅτι τῶν γη, ηθ δυο μέσαι ἀνάλογόν εἰσιν αἱ ζη, ηδ. ἤχθω γὰρ διὰ τοῦ ε τῇ αβ παράλληλος ἡ εκ· ἴση ἄρα ἐστὶν ἡ μὲν εκ τῇ ζη, ἡ δὲ κγ τῇ εδ. ἔσται γὰρ τοῦτο δῆλον ἀπὸ τοῦ λ ἐπὶ τὰ ε, ζ ἐπιζευχθεισῶν εὐθειῶν· ἴσαι γὰρ γίνονται αἱ ὑπὸ γλε, ζλδ, καὶ ὀρθαὶ αἱ πρὸς τοῖς κ, η· καὶ πάντα ἄρα πᾶσιν διὰ τὸ τὴν λε τῇ λζ ἴσην εἶναι· καὶ λοιπὴ ἄρα ἡ γκ τῇ ηδ ἴση ἐστίν. ἐπεὶ οὖν ἐστιν, ὡς ἡ δκ πρὸς κε, ἡ δη πρὸς ηθ, ἀλλ' ὡς ἡ δκ πρὸς κε, ἡ εκ πρὸς κγ· μέση γὰρ ἀνάλογον ἡ εκ τῶν δκ, κγ· ὡς ἄρα ἡ δκ

9 λε Bas., el B², αβ A

Therefore

$\Delta B : BE = BE : BA.$

But

$\Gamma B : B\Delta = \Delta B : BE.$

Therefore

$\Gamma B : B\Delta = B\Delta : BE = EB : BA.$

Q. E. F.

(iii) Heiberg III p. 66, 8 to 70, 5 (corresponds approximately to Diocles' Props. 11 to 13).

As Diocles in his book "On Burning Mirrors".

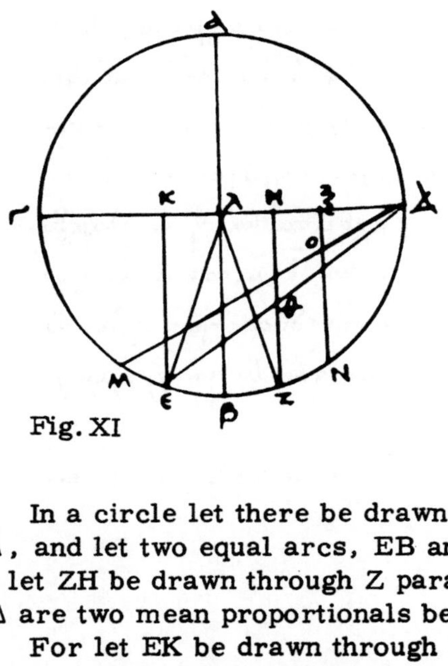

Fig. XI

In a circle let there be drawn two diameters at right angles, AB, ΓΔ, and let two equal arcs, EB and BZ, be cut off on either side of B; let ZH be drawn through Z parallel to AB; join ΔE. I say that ZH, HΔ are two mean proportionals between ΓH and HΘ.

For let EK be drawn through E parallel to AB. Then EK is equal to ZH, and KΓ to HΔ. (This will become clear if straight lines are drawn joining Λ to E and Z: for the angles ΓΛE, ZΛΔ are equal, and the angles at K and H are right; so all parts are equal to all [corresponding] parts, since ΛE equals ΛZ. Therefore the remainder [of ΓΛ-KΛ] ΓK equals HΔ [the remainder of ΛΔ-ΛH]). Then since

$\Delta H : H\Theta = \Delta K : KE,$

πρὸς κε καὶ ἡ εκ πρὸς κγ, οὕτως ἡ δη πρὸς ηθ. καί ἐστιν ἴση ἡ μὲν δκ τῇ γη, ἡ δὲ κε τῇ ζη, ἡ δὲ κγ τῇ ηδ. ὡς ἄρα ἡ γη πρὸς ηζ, ἡ ζη πρὸς ηδ καὶ ἡ δη πρὸς ηθ. ἐὰν δὴ παρ' ἑκάτερα τοῦ β ληφθῶσιν περιφέρειαι ἴσαι αἱ μβ, βν, καὶ διὰ μὲν τοῦ ν παράλληλος ἀχθῇ τῇ αβ ἡ νξ, ἐπιζευχθῇ δὲ ἡ δμ, ἔσονται πάλιν τῶν γζ, ξο μέσαι ἀνάλογον αἱ νξ, ξδ. πλειόνων οὖν οὕτως καὶ συνεχῶν παραλλήλων ἐκβληθεισῶν μεταξὺ τῶν β, δ, καὶ ταῖς ἀπολαμβανομέναις ὑπ' αὐτῶν περιφερείαις πρὸς τῷ β ἴσων τεθεισῶν ἀπὸ τοῦ β ὡς ἐπὶ τὸ γ, καὶ ἐπὶ τὰ γενάμενα σημεῖα ἐπιζευχθεισῶν εὐθειῶν ἀπὸ τοῦ δ, ὡς τῶν ὁμοίων ταῖς δε, δμ, τμηθήσονται αἱ παράλληλοι αἱ μεταξὺ τῶν β, δ κατά τινα σημεῖα, ⟨ὡς⟩ ἐπὶ τῆς προκειμένης καταγραφῆς τὰ ο, θ, ἐφ' ἃ κανόνος παραθέσει ἐπιζεύξαντες εὐθείας ἕξομεν καταγεγραμμένην ἐν τῷ κύκλῳ τινὰ γραμμήν, ἐφ' ἧς ἐὰν ληφθῇ τυχὸν σημεῖον καὶ δι' αὐτοῦ παράλληλος ἀχθῇ τῇ λβ, ἔσται ἡ ἀχθεῖσα καὶ ἡ ἀπολαμβανομένη ὑπ' αὐτῆς ἀπὸ τῆς διαμέτρου πρὸς τῷ δ μέσαι ἀνάλογον τῆς τε ἀπολαμβανομένης ὑπ' αὐτῆς ἀπὸ τῆς διαμέτρου πρὸς τῷ γ σημείῳ καὶ τοῦ μέρους αὐτῆς τοῦ ἀπὸ τοῦ ἐν τῇ γραμμῇ σημείου ἐπὶ τὴν γδ διάμετρον.

τούτων προκατεσκευασμένων ἔστωσαν αἱ δοθεῖσαι δύο εὐ-

23-24 ὡς ἐπὶ Bas., B ("ut in"), επι A

24-25 παραθέσει B ("appositione"), παραθεσεις A

30 τοῦ ἐν Bas., τῆς ἐν EGH², τη εν A

32 προκατεσκευασμένων H, προκατασκευασμενων A

but ΔK : KE = EK : KΓ (for EK is the mean proportional between ΔK and KΓ),
it follows that

ΔK : KE = EK : KΓ = ΔH : Hθ.

Now

ΔK = ΓH, KE = ZH, and KΓ = HΔ.

Therefore

ΓH : HZ = ZH : HΔ = ΔH : Hθ.

So if equal arcs MB and BN are taken on either side of B, and NΞ is drawn through N parallel to AB, and ΔM is joined, NΞ and ΞΔ will again be [two] mean proportionals between ΓΞ and ΞO. So if more parallels [to AB] are drawn in this manner in close succession to another between B and Δ, and arcs equal to the arcs cut off by these parallels from B are marked off on the other side of B towards Γ, and straight lines are drawn from Δ to the points thus generated, like ΔE and ΔM, then the parallels [to AB] between B and Δ will be cut at certain points, like O and θ in the present figure. If we place our ruler against these points and join them by straight lines, we will have a certain line drawn in the circle. Then if we take an arbitrary point on that line and draw through it a line parallel to ΛB, the parallel and the section of the diameter it cuts off from Δ will be [two] mean proportionals between the section of the diameter it [the parallel] cuts off from Γ, and that part of it [the parallel] between the point on the line [the "cissoid"] and the diameter ΓΔ.

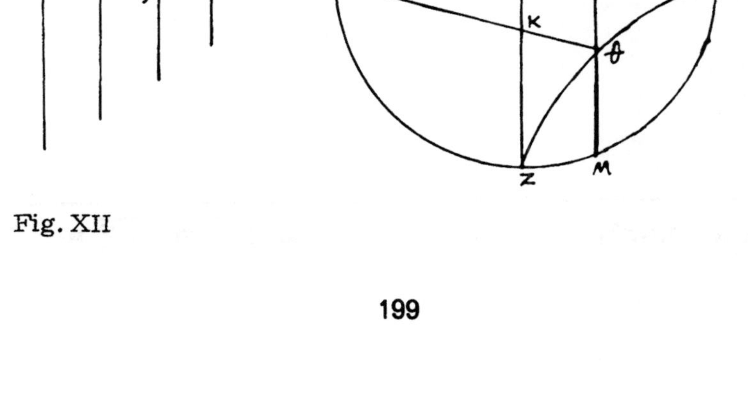

Fig. XII

θεῖαι, ὧν δεῖ δύο μέσας ἀνάλογον εὑρεῖν, αἱ α̅, β̅, καὶ ἔστω κύκλος, ἐν ᾧ δύο διάμετροι πρὸς ὀρθὰς ἀλλήλαις αἱ γ̅δ̅, ε̅ζ̅, καὶ γεγράφθω ἐν αὐτῇ ἡ διὰ τῶν συνεχῶν σημείων γραμμή, ὡς προείρηται, ἡ δ̅θ̅ζ̅, καὶ γεγονέτω, ὡς ἡ α̅ πρὸς τὴν β̅, ἡ γ̅η̅ πρὸς η̅κ̅, καὶ ἐπιζευχθεῖσα ἡ γ̅κ̅ καὶ ἐκβληθεῖσα τεμνέτω τὴν γραμμὴν κατὰ τὸ θ̅, καὶ διὰ τοῦ θ̅ τῇ ε̅ζ̅ παράλληλος ἤχθω ἡ λ̅μ̅. διὰ ἄρα τὰ προγεγραμμένα τῶν γ̅λ̅, λ̅θ̅ μέσαι ἀνάλογόν εἰσιν αἱ μ̅λ̅, λ̅δ̅. καὶ ἐπεί ἐστιν, ὡς ἡ γ̅λ̅ πρὸς λ̅θ̅, οὕτως ἡ γ̅η̅ πρὸς η̅κ̅, ὡς δὲ ἡ γ̅η̅ πρὸς η̅κ̅, οὕτως ἡ α̅ πρὸς τὴν β̅, ἐὰν ἐν τῷ αὐτῷ λόγῳ ταῖς γ̅λ̅, λ̅μ̅, λ̅δ̅, λ̅θ̅ παρεμβάλωμεν μέσας τῶν α̅, β̅, ὡς τὰς ν̅, ξ̅, ἔσονται εἰλημμέναι τῶν α̅, β̅ μέσαι ἀνάλογον αἱ ν̅, ξ̅· ὅπερ ἔδει εὑρεῖν.

---

34 διάμετροι BGE²H², διαμετρον A   36 δ̅θ̅ζ̅ B, δ̅ε̅ζ̅ A

Now that we have made this preliminary construction, let the two straight lines between which we must find two mean proportionals be A and B. Let there be a circle, in which there are two diameters, ΓΔ, EZ, at right angles to each other, and let there be drawn in it the line generated from the continuous points, as described above, ΔΘZ.

Let ΓH : HK = A : B,

join ΓK and produce it to cut the line at Θ. Let ΛM be drawn through Θ parallel to EZ. Then, from the above, MΛ and ΛΔ are [two] mean proportionals between ΓΛ and ΛΘ. And since

ΓH : HK = ΓΛ : ΛΘ and A : B = ΓH : HK,

if we construct between A and B proportionals N and Ξ in the same ratio as ΓΛ : ΛM : ΛΔ : ΛΘ [i. e. make

A : N : Ξ : B = ΓΛ : ΛM : ΛΔ : ΛΘ],

N and Ξ will have been found as two mean proportionals between A and B.

Q. E. F.

Appendix B

Other ancient and medieval proofs of the focal property of the parabola.

The lettering of the figures of these proofs has been changed from the originals to conform with that of Diocles' Prop. 1, to facilitate comparison.

(i) The Bobbio Mathematical Fragment (Heiberg, *Mathematici Graeci Minores* pp. 87-88).

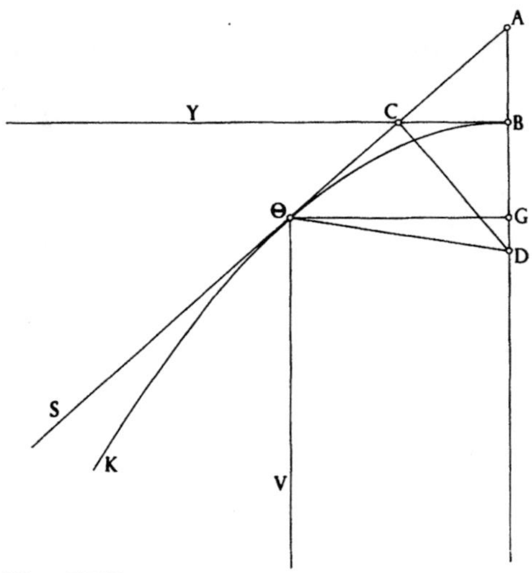

Fig. XIII

See Fig. XIII. Y is the parameter, D a point on the axis such that $BD = \frac{1}{4}Y$. Then

$Y.BG = \theta G^2$ (property of the parabola).

But

$Y = 4\,BD$ (construction).

$\therefore\ 4\,BD \,.\, BG = \theta G^2.$

But

$BG = BA$ [1]

and

$\theta G = 2\,BC$ (similar triangles),

$\therefore\ BD \,.\, BA = BC^2.$

$\therefore\ \hat{C}$ is right.

And since $AC = C\theta,\ AD = D\theta,$

$\therefore\ \widehat{DA\theta} = \widehat{D\theta A}$

⟨But

$\widehat{V\theta S} = \widehat{DA\theta},$

$\therefore\ \widehat{D\theta A} = \widehat{V\theta S}.$⟩ [2]

But

$\widehat{A\theta B} = \widehat{S\theta K},$

$\therefore\ \widehat{D\theta B} = \widehat{K\theta V}.$ [3]

(ii) Ibn al-Haytham, "On Parabolic Burning Mirrors" (text, *Majmūʿ*, fourth risāla, pp. 4-5; translation, Heiberg-Wiedemann, *Bibliotheca Mathematica* X p. 209).

See Fig. XIV. Y is the parameter, D a point on the axis such that $BD = \frac{1}{4}Y.$

$AB = BG$, because $S\theta A$ is a tangent and θG an ordinate [4],

$\therefore 4\,DB \,.\, BG + GD^2 = AD^2.$ [5]

But

$BD = \frac{1}{4}Y.$

$\therefore 4\,BD \,.\, BG = Y \,.\, BG.$

$\therefore Y \,.\, BG + GD^2 = AD^2.$

---

1 Cf. Diocles § 40.

2 The section in brackets is not in the text, but it is an essential part of the proof.

3 On this use of "mixed" angles see note on § 81.

4 Cf. Diocles § 40.

5 By Euclid II 8, $4DB.BG + GD^2 = (DB + BG)^2$; since $BG = AB$, $(DB + BG)^2 = AD^2$. Cf. Diocles' procedure, second note on § 118.

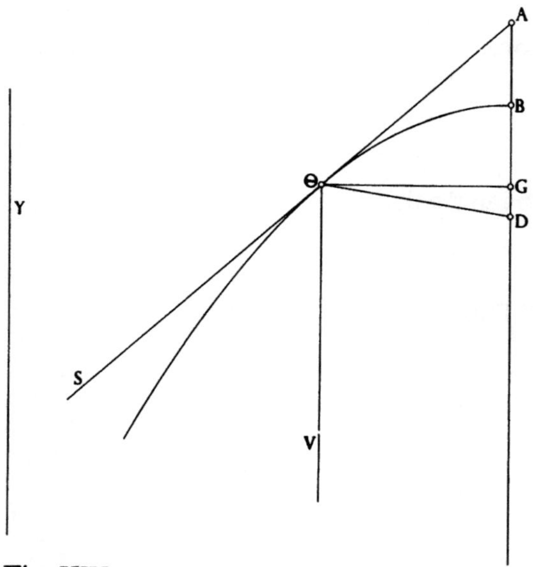

Fig. XIV

But

$Y . BG = \theta G^2$ (property of the parabola),

$\therefore \theta G^2 + DG^2 = AD^2$.

But

$\theta G^2 + DG^2 = \theta D^2$ ($\widehat{\theta G D}$ a right angle)

$\therefore AD^2 = \theta D^2$.

$\therefore AD = \theta D$.

$\therefore \widehat{DA\theta} = \widehat{D\theta A}$.

But

$\widehat{V\theta S} = \widehat{DA\theta}$ ($V\theta \parallel DA$),

$\therefore \widehat{D\theta A} = \widehat{V\theta S}$.

Appendix C

Analysis of Archimedes' problem and Diocles' solution, by O. Neugebauer.

(i) Archimedes, *Sphere and Cylinder* II 2.

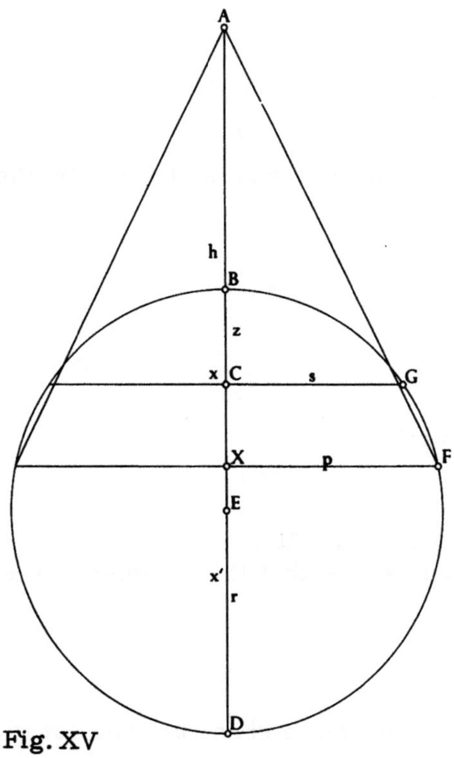

Fig. XV

See Fig. XV, in which AX = h, BX = x, XD = x', BC = z, ED = r, CG = s, XF = p.

The volume S of a segment of a sphere of radius r and height x equals the volume of a cone of the same base and height h if

$$\frac{h}{x} = \frac{r + x'}{x'} \quad (x + x' = 2r). \tag{1}$$

A modern proof is as follows.

$$s^2 = z(2r - z).$$

Therefore

$$S = \pi \int_0^x z(2r - z)\,dz = \pi \left\{ rz^2 - \frac{1}{3} z^3 \right\}_0^x$$

$$= \pi x^2 (r - \frac{x}{3}) = \frac{\pi}{3} x^2 (3r - x).$$

Since

$$x + x' = 2r,$$

$$S = \frac{\pi}{3} x^2 (3r - x) = \frac{\pi}{3} x^2 (r + x').$$

The cone of volume S with base of radius p and height h must therefore satisfy:

$$\frac{\pi}{3} p^2 h = \frac{\pi}{3} x^2 (r + x'),$$

or, since

$$p^2 = xx',$$

$$xx'h = x^2 (r + x'),$$

thus $\frac{h}{x} = \frac{r + x'}{x'}$, which is (1).

(ii) Archimedes, *Sphere and Cylinder* II 4.

To cut a given sphere of radius r so that the volumes of the segments have a given ratio:

$$S_1 : S_2 = \lambda : \mu.$$

According to the preceding (II 2) one has to construct two cones of altitude h and h' such that

$$\frac{h}{x} = \frac{r + x'}{x'}, \quad \frac{h'}{x'} = \frac{r + x}{x} \tag{2}$$

and

$$\frac{h}{h'} = \frac{\lambda}{\mu}, \tag{3}$$

because $S_1 : S_2 = h : h'$, since the cones have a common base.

[*Note:* Since

$$\frac{3}{\pi} S_2 = x^2 (3r - x) = 3rx^2 - x^3,$$

and

$$\frac{3}{\pi} S_1 = x'^2 (3r - x') = (2r - x)^2 (r + x)$$

$$= 4r^3 - 3rx^2 + x^3 = 4r^3 - \frac{3}{\pi} S_2,$$

for the volume S of the sphere one has

$$\frac{3}{\pi}(S_1 + S_2) = 4r^3 = \frac{3}{\pi} S,$$ which is correct.]

From

$$\lambda S_2 = \mu S_1 = \lambda (S - S_1)$$

one has

$$\lambda (3rx^2 - x^3) = \mu (4r^3 - 3rx^2 + x^3),$$

or

$$x^3 (\lambda + \mu) - 3r(\lambda + \mu) x^2 + 4\mu r^3 = 0.$$

Thus for x one has the cubic equation

$$x^3 - 3rx^2 + \frac{4\mu r^3}{\lambda + \mu} = 0,$$

which is equivalent to (10) and (11) below, derived by Archimedes.

Transformation of (2) and (3) to a cubic equation:

From (2)

$$\frac{h - x}{x} = \frac{r}{x'}, \quad \frac{h' - x'}{x'} = \frac{r}{x},$$

hence

$$\frac{r}{h - x} = \frac{x'}{x} = \frac{h' - x'}{r}, \tag{4}$$

or

$$\frac{r}{h' - x'} = \frac{h - x}{r},$$

thus

$$\frac{h' - x' + r}{h' - x'} = \frac{h - x + r}{r},$$

therefore

$$\frac{r}{h' - x'} = \frac{h - x + r}{h' - x' + r'}\text{, whence, adding 1 to both sides,}$$

$$\frac{h' - x' + r}{h' - x'} = \frac{h' + h - (x + x') + 2r}{h' - x' + r} = \frac{h' + h}{h' - x' + r} \quad (x + x' = 2r).$$

Therefore

$$(h' - x' + r)^2 = (h' + h)(h' - x'),$$

or

$$\frac{h' + h}{h' - x'} = \frac{(h' - x' + r)^2}{(h' - x')^2}. \tag{5}$$

From (4)

$$\frac{r}{h' - x'} = \frac{x}{x'}, \text{ or } \frac{h' - x' + r}{h' - x'} = \frac{x + x'}{x'} = \frac{2r}{x'}. \tag{6}$$

From (3)

$$\frac{h + h'}{h'} = \frac{\lambda + \mu}{\mu}, \tag{7}$$

which is a given ratio. But

$$\frac{h' + h}{h'} = \frac{h' + h}{h' - x'} \cdot \frac{h' - x'}{h'}, \tag{8}$$

and, from (5) and (6),

$$\frac{h' + h}{h' - x'} = \frac{(2r)^2}{x'^2},$$

and, from (2),

$$\frac{h' - x'}{h'} = 1 - \frac{x'}{h'} = 1 - \frac{x}{r + x} = \frac{r}{r + x}.$$

Hence, from (8) and (7),

$$\frac{h' + h}{h'} = \frac{(2r)^2}{x'^2} \cdot \frac{r}{r + x} = \frac{\lambda + \mu}{\mu}. \tag{9}$$

If we define c by

$$\frac{r}{c} = \frac{h' + h}{h'} = \frac{\lambda + \mu}{\mu}, \text{ or } c = r \cdot \frac{\mu}{\lambda + \mu} \text{ or } \frac{r - c}{c} = \frac{\lambda}{\mu}, \tag{10}$$

then c is a given *parameter*.

Then, from (9) and (10)

$$\frac{r}{c} = \frac{(2r)^2}{x'^2} \cdot \frac{r}{r + x},$$

or

$$\frac{r+x}{c} = \frac{(2r)^2}{x'^2},$$

and, since $x' = 2r - x$, the cubic equation for x is

$$\frac{r+x}{c} = \frac{(2r)^2}{(2r-x)^2}. \qquad (11)$$

Geometric interpretation:

See Fig. XVI. Given $BZ = r$, $BD = 2r$, $Z\theta = c < r$, find a point X such that

$$\frac{XZ}{Z\theta} = \frac{BD^2}{DX^2}.$$

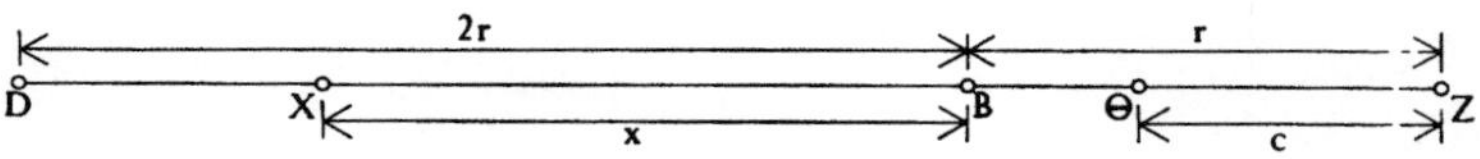

Fig. XVI

(iii) Diocles' solution.

See Fig. XVII. Given: $\frac{\lambda}{\mu}$, $AB = 2r$, $AH = r$, construct the rectangle $AH\theta B$, where

$AL = AH = r$,
$BM = B\theta = r$,
$RM = r$.
Draw RBQ.

We call RQ ξ and define η from

$$\frac{\xi}{\eta} = \frac{\lambda}{2\mu}. \qquad (12)$$

We then construct an ellipse with diameter $RQ = \xi$, conjugate direction $LQ \parallel RM$, and parameter η from (12). Then it is possible to construct a hyperbola with $H\theta$ and HA as asymptotes, passing through the point B. There must be a point S where it intersects the ellipse. Since B and S are points of the hyperbola one has

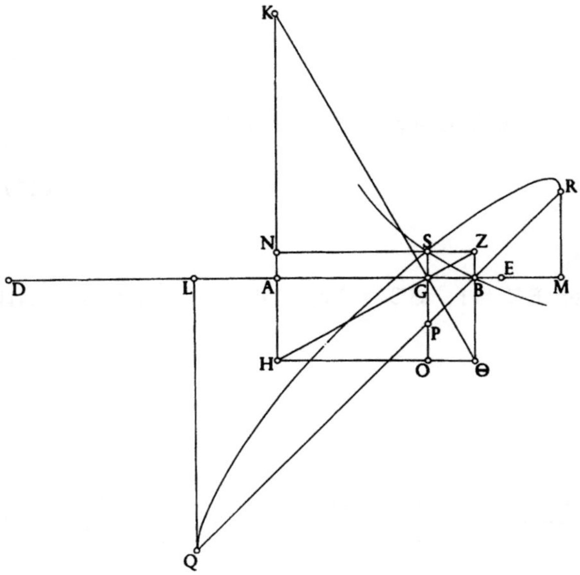

Fig. XVII

$NS \cdot SO = AB \cdot B\theta$ (NH, Hθ asymptotes),

and the intersection G of SO and AB must be a point of diagonal HZ. Having so defined the point G we call

$AG = x', \quad GB = x,$

and draw θGK. Furthermore, we define D and E from

$KA = AD, \quad ZB = BE.$

Since S is a point of the ellipse its coordinates

$y = SP, \quad z = PR$

must, because of (12), satisfy the equation

$$\frac{z(\xi - z)}{y^2} = \frac{\eta}{\xi} = \frac{2\mu}{\lambda}. \tag{13}$$

Furthermore, by construction

$$\frac{RB}{BP} = \frac{BM}{x} = \frac{r}{x}, \text{ hence } \frac{RP}{BP} = \frac{r+x}{x}.$$

Since

$$\frac{BP}{PQ} = \frac{x}{LG} = \frac{x}{r+x'},$$

we have

$$\frac{RP}{PQ} = \frac{z}{\xi - z} = \frac{r+x}{r+x'},$$

thus

$$\frac{z(\xi - z)}{(\xi - z)^2} = \frac{(r+x)(r+x')}{(r+x')^2},$$

or

$$\frac{z(\xi - z)}{(r+x)(r+x')} = \frac{(\xi - z)^2}{(r-x')^2} = \frac{PQ^2}{LG^2} = 2.$$

Thus we have

$$z(\xi - z) = 2(r+x)(r+x'),$$

and, from (13),

$$z(\xi - z) = \frac{2\mu}{\lambda} y^2.$$

Hence

$$\frac{\mu}{\lambda} = \frac{(r+x)(r+x')}{y^2}.$$

But $y = SP = PG + BZ = x + BE = GE$.
Thus finally

$$\frac{\mu}{\lambda} = \frac{(r+x)(r+x')}{GE^2}. \tag{14}$$

Since

$$\frac{KA}{\theta B} = \frac{x'}{x} = \frac{HA}{ZB},$$

we have also

$$\frac{KA + x'}{\theta B + x} = \frac{x'}{x} = \frac{HA + x'}{ZB + x},$$

or $(KA + x')(ZB + x) = (\theta B + x)(HA + x')$,
i.e. $DG.GE = (r+x)(r+x')$.
Therefore, from (14),

$$\frac{\mu}{\lambda} = \frac{(r+x)(r+x')}{GE^2} = \frac{DG.GE}{GE^2} = \frac{DG}{GE}. \tag{15}$$

Furthermore, by construction,

$$\frac{\theta B}{x} = \frac{KA}{x'} = \frac{DA}{x'}, \quad \frac{HA}{x'} = \frac{BZ}{x} = \frac{EB}{x},$$

or

$$\frac{r}{x} = \frac{DG - x'}{x'}, \quad \frac{r}{x'} = \frac{GE - x}{x}. \qquad (16)$$

Thus (15) and (16) imply, if we call GE = h, DG = h',

$$\frac{h}{h'} = \frac{\lambda}{\mu}, \quad \frac{r}{x} = \frac{h' - x'}{x'}, \quad \frac{r}{x} = \frac{h - x}{x},$$

which are the conditions for h, h', x, and x' (= 2 r - x) corresponding to a ratio $S_1 : S_2 = \lambda : \mu$ for the volumes of the segments (cf. (2) and (3) above).

Appendix D

Proof that rays reflected from the surface formed by the revolution of a parabola about a chord perpendicular to the axis do not all meet the circumference of a circle, by O. Neugebauer

(i) Definitions and terminology.

We consider two planes at right angles to each other, which we call "horizon" and "meridian". Their intersection is the "axis". All "incident rays" are parallel to the axis.

On occasion we use the terms "parabola" and "circle" for brevity's sake, when in fact only certain arcs of these curves are considered.

The "reflecting surface" (or the "mirror") is generated by rotating a parabola in the meridian plane by $\pm 90°$ around the straight line AB (see Fig. XVIII for the meridian plane, Figs. XIX and XX for the horizon).

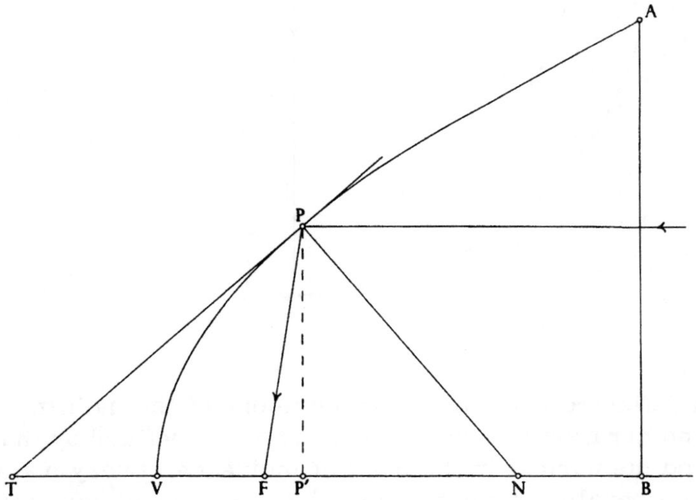

Fig. XVIII

(ii) Consider an incident ray in the meridian plane which meets the parabola in P. Its reflected ray is PF, making with the normal PN the same angle as the incident ray. Hence the focus F is a point of the curve in which reflected rays hit the horizon.

We now consider all incident rays lying in a plane which is parallel to the horizon and contains the point P. These rays meet the mirror in a circle of radius BP' (cf. Fig. XIX). The normals to the reflecting surface in this circle form a right circular cone which meets the horizon in a circle HNG.

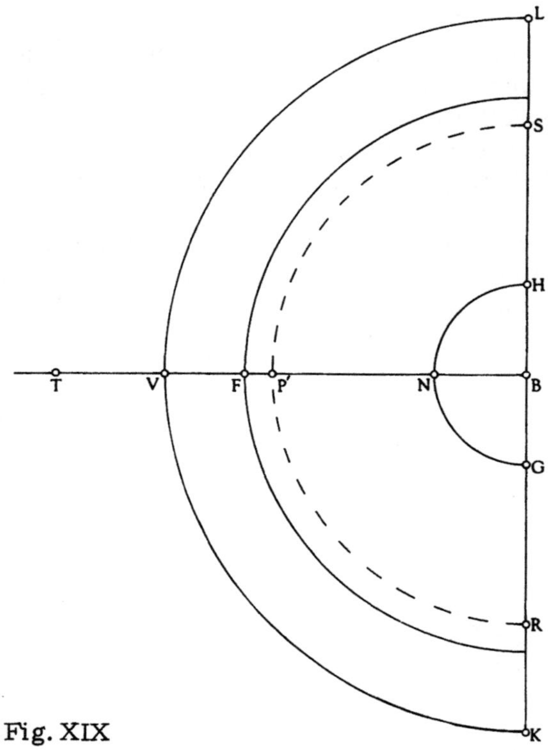

Fig. XIX

(iii) Each reflected ray which belongs to one of the incident rays under consideration lies in the plane which is defined by that incident ray and the normal in the point of incidence. Every plane which contains a parallel to the horizon, i.e. an incident ray, intersects the horizon in a straight line parallel to the axis. Therefore a ray reflected at a point Q of the mirror (cf. Fig. XX) must meet the horizon in a point of this straight line, one point of which is the point M, where the normal QM meets the horizon. Thus the reflected ray meets the horizon in some point Z of this line such that it makes at Q (not at Q', which is the projection of Q on to the horizon) equal angles with the normal QM.

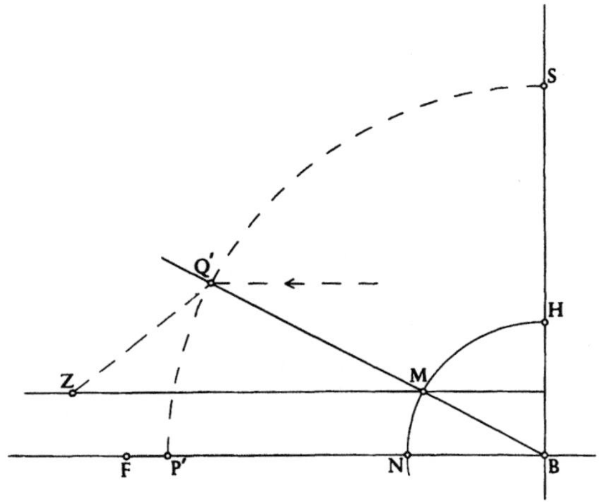

Fig. XX

(iv) Hence we see that the curve in which the reflected rays meet the horizon originates from the points of a circle HNG by the addition of the parallel vectors MZ (cf. Fig. XX). One of these vectors has the length NF. The endpoints of these parallel vectors do not lie on a circle, since the vectors increase in length as Q moves towards ±90° from P; for then the angle between incident ray and normal tends toward 90°, hence the angle between normal and reflected ray also tends toward 90°, therefore the reflected ray tends to become parallel to the axis, and thus meets the horizon at ever increasing distances. Hence not all reflected rays meet the horizon in a circle.

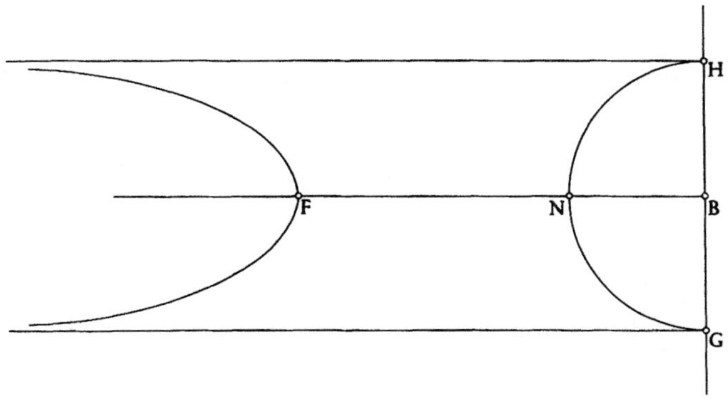

Fig. XXI

(v) The curve in question must look like that shown in Fig. XXI. As the plane of the incident rays varies its distance from the horizon the radius of the circle HNG varies and so does the width of the curve in which the reflected rays meet the horizon. All these curves have only the point F in common and will fill an area from the axis to the width of the mirror.

Bibliography

Afshar, Iraj: Bibliographie des Catalogues des Manuscrits Persans. Tehran: 1958.

Almagest: see Ptolemy.

Apollonius: Apollonii Pergaei quae Graece exstant cum commentariis Eutocii (ed. J. L. Heiberg), 2 vols. Leipzig: 1891, 1893.

Arberry, A. J.: The Chester Beatty Library, A Handlist of the Arabic Manuscripts, Vol. VII. Dublin: 1964.

Archimedes: Archimedis Opera Omnia cum commentariis Eutocii, (iterum ed. J. L. Heiberg), 3 vols. Leipzig: 1910-1915.

Archimedes: see also Heath.

Aristarchus of Samos: On the Sizes and Distances of the Sun and Moon (ed. T. Heath). Oxford: 1913.

Aristotle, Nicomachean Ethics: Aristotelis Ethica Nicomachea (ed. I. Bywater). Oxford: 1894.

Aristotle, Prior Analytics: Aristotelis Analytica Priora et Posteriora (ed. W. D. Ross and L. Minio-Paluello). Oxford: 1964.

Autolycus: J. Mogenet, Autolycus de Pitane. Louvain, 1950 (Université de Louvain, Recueil de Travaux d'Histoire et de Philologie, 3e. Série Fasc. 37).

Awad, Gurgis: "Arabic Manuscripts in American Libraries". Sumer 7, 237-277 (1951). (Arabic).

Bachmann, Peter: Galens Abhandlung darüber, daß der vorzügliche Arzt Philosoph sein muß. Göttingen: 1965 (Ak. Wiss. Göttingen, Nachrichten Phil.-hist. Kl. 1965. 1).

Belger, C.: "Ein neues Fragmentum Mathematicum Bobiense". Hermes 16, 261-84 (1881).

Boilot, D. J.: "L'oeuvre d'al-Beruni, essai bibliographique". Mélanges de l'Institut Dominicain d'Etudes Orientales du Caire 2, 161-256 (1955).

Bretschneider, C. A.: Die Geometrie und die Geometer vor Eukleides. Leipzig: 1870.

Brockelmann, Carl: Geschichte der Arabischen Litteratur, zweite den Supplementbänden angepasste Auflage, 2 vols. Leiden: 1943, 1949 [GAL] [and] Supplementbände I-III. Leiden: 1937, 1938, 1942 [S].

Bulmer-Thomas, I.: "Conon of Samos". Dictionary of Scientific Biography III, (New York), 391 (1971).

Bulmer-Thomas, I.: "Dionysodorus". Dictionary of Scientific Biography IV (New York), 108-10 (1971).

Caley, Earle R.: Orichalcum and Related Ancient Alloys. New York: 1964 (Numismatic Notes and Monographs, 151).

Cantor, Moritz: "Über das neue Fragmentum Mathematicum Bobiense". Hermes 16, 640-42 (1881).

Cantor, Moritz: Vorlesungen über Geschichte der Mathematik. Vol. I, 3rd edn. Leipzig and Berlin: 1907.

Casiri, M.: Bibliotheca Arabico-Hispana Escurialensis, 2 vols. Madrid: 1760, 1770.

CLA: see Lowe.

Crönert, Wilhelm: "Der Epikureer Philonides". SB Berlin 1900.2 942-59.

Daiber, Hans: Die Arabische Übersetzung der Placita Philosophorum. Inaugural-Dissertation, Saarbrücken, 1968.

Datta, Bimal Kumar: Libraries and Librarianship of Ancient and Medieval India. Delhi: 1970.

Derenbourg, H., Renaud, H. P. -J.: Les Manuscrits Arabes de l'Escurial, Vol. II.2. Paris: 1941.

Descartes, René: Géométrie, in D. E. Smith and M. L. Latham, The Geometry of René Descartes. Chicago and London: 1925 [with facsimile of the original text, pp. 297-412 of his Discours de la Méthode, 1637].

Diels, Hermann: Antike Technik, 2nd edn. Leipzig and Berlin: 1920.

Dijksterhuis, E. J.: Archimedes. Copenhagen: 1956.

Drecker, Joseph: Die Theorie der Sonnenuhren. Berlin and Leipzig: 1925 (Die Geschichte der Zeitmessung und der Uhren, Bd. I, Lf. E).

Etudes Thasiennes V: Recherches sur l'histoire et les cultes de Thasos II, par Christiane Dunant et Jean Pouilloux. Paris: 1958.

Euclid: The Thirteen Books of Euclid's Elements, tr. T. L. Heath, 2nd edn., 3 vols. Cambridge: 1926.

Euclid, Phaenomena: Euclids Phaenomena et Scripta Musica (ed. H. Menge). Leipzig: 1916 (Euclidis Opera Omnia VIII).

al-Farabi: Matematicheskie Traktate (ed. S. E. Esenov). Alma-Ata: 1972 [Russian translations by various hands].

Fermat, Pierre de: Oeuvres de Fermat (ed. P. Tannery and Ch. Henry), 4 vols. Paris:1891-1912.

Forbes, R. J.: Studies in Ancient Technology, Vol. VIII, 2nd edn. Leiden: 1971.

Fraser, P. M.: Ptolemaic Alexandria, 3 vols. Oxford: 1972.

GAL: see Brockelmann.

Galen: Claudii Galeni Opera Omnia (ed. C. G. Kühn), 20 vols. Leipzig: 1821-33.

Galen: Scripta Minora (ed. I. Mueller), Vol. II. Leipzig: 1891.

Galen: Compendium Timaei Platonis (ed. P. Kraus and R. Walzer). London: 1951 (Plato Arabus I).

Galen: On the Parts of Medicine On Cohesive Causes, etc. (ed. Malcolm Lyons). Berlin: 1969 (Corpus Medicorum Graecorum, Supplementum Orientale II).

Galen: see also Bachmann.

Gandz, Solomon: "Der Hultsch-Cantorsche Beweis von der Reihenfolge der Buchstaben in den mathematischen Figuren der Griechen und Araber". Quellen und Studien zur Geschichte der Mathematik, Astronomie und Physik, B 2, 81-97 (1933).

GAS: see Sezgin.

Georr, Khalil: Les Catêgories d'Aristote dans leurs versions syro-arabes. Beirut: 1948.

Gibbs, Sharon: Greek and Roman Sundials. Ph. D. Dissertation, Yale University, 1972 (University Microfilms Dissertation #73-14334).

Golchīn-Maʿānī, Ahmad: Fihrist Kutub Kaṭṭī Kitābkāna-i Āstān-i Quds Riḍawī, Vol. 8. Meshhed: 1350 [1972] [catalogue of the mathematical and astronomical manuscripts of the Shrine Library].

Gomes Teixeira, F.: Traitê des Courbes Spêciales Remarquables Planes et Gauches, Vol. 1. Coimbra: 1908 (Gomes Teixeira, Obras sobre Mathematica, Vol. 4).

Handlist: see Voorhoeve.

Haschmi, Mohamed Y.: Propagations of Ray...from...al-Kindi. Aleppo: 1967.

Heath, T. L.: The Works of Archimedes. Cambridge: 1897.

Heath, T. L.: A History of Greek Mathematics, 2 vols. Oxford: 1921 [HGM].

Heiberg, J. L.: "Zum Fragmentum mathematicum Bobiense". Zeitschrift für Mathematik und Physik 28, (Hist.-lit. Abt.), 121-29 (1883)

Heiberg, J. L.: "Ptolemäus de Analemmate". Abhandlungen zur Geschichte der Mathematik VII, 3-30 (1895) (= Zeitschrift für Mathematik und Physik 40, Supp.).

Heiberg, J. L.: "Eine neue Archimedeshandschrift". Hermes 42, 235-303 (1907).
Heiberg, J. L.: see also Apollonius; Archimedes; [Heron]; Mathematici.
Heiberg, J. L., Wiedemann, E.: "Ibn al Haiṯams Schrift über parabolische Hohlspiegel". Bibliotheca Mathematica 3 F. X, 201-37 (1910).
Heron: Pneumatica et Automata (ed. W. Schmidt) Leipzig: 1899 (Heronis Alexandrini Opera quae supersunt omnia I).
Heron: Mechanica et Catoptrica (ed. L. Nix and W. Schmidt). Leipzig: 1900 (Heronis Alexandrini Opera quae supersunt omnia II. 1).
[Heron]: Geometrica (ed. J. L. Heiberg). Leipzig: 1912 (Heronis Alexandrini Opera quae supersunt omnia IV).
HGM: see Heath.
Hippolytus: Refutatio omnium haeresium (ed. Paul Wendland). Leipzig: 1916 (Die Griechischen Christlichen Schriftsteller der ersten drei Jahrhunderte: Hippolytus, Werke, 3).
Huxley, G. L.: Anthemius of Tralles, A Study in Later Greek Geometry. Cambridge, Mass.: 1959.
Hultsch, F.: "Dositheos 9". Paulys Real Encyclopädie der Classischen Altertumswissenschaften, new edition by G. Wissowa, V. 2. cols. 1607-08. Stuttgart: 1905, [RE].
Ibn Abī Uṣaybiʿa: Kitāb ʿuyūn al-anbāʾ fi ṭabaqāt al-aṭibbāʾ, (ed. A. Müller). 2 vols. Cairo: 1882.
Ibn al-Haytham: Majmūʿ al-rasāʾil. Hyderabad-Deccan, 1357 A. H. [= A. D. 1938/9] (Osmania Oriental Publications Bureau).
Ibn al-Qifti: Taʾriḵ al-Ḥukamāʾ, (ed. J. Lippert). Leipzig: 1903.
IG II^2: Inscriptiones Graecae Voluminis II et III Editio Minor, Pars Altera Fasc. I (ed. J. Kirchner). Berlin: 1927.
IGR: Inscriptiones Graecae ad res Romanas pertinentes, (ed. R. Cagnat and G. Lafaye), Vol. IV. Paris: 1906.
Kepler, J.: Ad Vitellionem Paralipomena (Astronomiae Pars Optica) (ed. Franz Hammer). München: 1939 (Gesammelte Werke II).
Kisa, Anton: Das Glas im Altertume, 3 vols. Leipzig: 1908 (Hiersemanns Handbücher III).
Krasnova, S. A.: "Abu-l-Vafa al-Buzdzhani, Kniga o tom, chto neobkhodimo remeslenniku iz geometricheskikh postroeniy". Fiziko-matimaticheskie nauki v stranakh vostoka I(IV). Moscow: 1966, pp. 42-140.
Krause, Max: "Stambuler Handschriften islamischer Mathematiker". Quellen und Studien zur Geschichte der Mathematik, Astronomie und Physik, B 3, 437-532 (1936).

Kunitzsch, Paul: Der Almagest, Die Syntaxis Mathematica des Claudius Ptolemäus in arabisch-lateinischer Überlieferung. Wiesbaden: 1974.

Lejeune, Albert: Recherches sur la catoptrique grecque. Bruxelles: 1957 (Académie Royale de Belgique, Mémoires LII. 2).

Letronne, A. J.: "Mémoire sur l'utilité qu'on peut retirer de l'étude des noms propres grecs pour l'histoire et l'archéologie". Letronne, Oeuvres Choisies (ed. E. Fagnan), 3 Sér., Vol. 2. Paris: 1885, pp. 1-126.

Loria, G.: Spezielle Algebraische und Transzendente Ebene Kurven, 2nd edn., 2 vols. Leipzig and Berlin: 1910.

Lowe, E. A.: Codices Latini Antiquiores, Vol. III. Oxford: 1938 [CLA].

LSJ: A Greek-English Lexicon compiled by H. G. Liddell and R. Scott. A new Edition... by H. Stuart Jones. Oxford: 1940.

Maass, E.: Aratea. Berlin: 1892.

Mathematici Graeci Minores: (ed. J. L. Heiberg). Copenhagen: 1927 (Det Kgl. Danske Vidensk. Selsk., Hist.-fil. Meddelelser XIII, 3).

Molther, Johannes: Problema Deliacum de Cubi Duplicatione. Frankfurt: 1619.

Montucla, J. F.: Histoire des Mathématiques, 2nd edn., Vol. I. Paris: An VII =[1798/9].

Mugler, Charles: Dictionnaire historique de la terminologie géométrique des Grecs. Paris: 1958.

Neugebauer, O.: "Apollonius-Studien". Quellen und Studien zur Geschichte der Mathematik, Astronomie und Physik, B 2, 215-54 (1933).

Neugebauer, O.: A History of Ancient Mathematical Astronomy, 3 vols. New York: 1975.

Neugebauer, O., van Hoesen, H. B.: Greek Horoscopes. Philadelphia: 1959 (American Philosophical Society, Memoirs, 48).

Nix, L.: Das Fünfte Buch der Conica des Apollonius von Perga. Leipzig: 1889.

Pappus, Collection: Pappi Alexandrini Collectionis quae supersunt (ed. F. Hultsch), 3 vols. Berlin: 1875-78.

Pliny: Plinius, Historia Naturalis, Bk. II (ed. Jean Beaujeu). Paris: 1950.

Proclus Diadochus: Hypotyposis Astronomicarum Positionum (ed. K. Manitius). Leipzig: 1909.

Proclus Diadochus: In Primum Euclidis Elementorum Librum Commentarii (ed. G. Friedlein). Leipzig: 1873.

Proclus Diadochus: Commentariorum Libri IIII. A Francisco Barocio primum iam Romanae linguae venustate donati. Padua. 1560.

Ptolemy, Almagest: Syntaxis Mathematica (ed. J. L. Heiberg), 2 vols. Leipzig: 1889, 1903 (Claudii Ptolemaei Opera quae exstant omnia I).
Ptolemy: Opera Astronomica Minora (ed. J. L. Heiberg). Leipzig: 1907 (Claudii Ptolemaei Opera quae exstant omnia II).
Ptolemy: The Arabic Version of Ptolemy's Planetary Hypotheses (ed. Bernard R. Goldstein). Philadelphia: 1967 (Transactions of the American Philosophical Society 57. 4).
Rehm, A.: "Konon 11" Paulys Real-Encyclopädie der Classischen Altertumswissenschaft, new edition by G. Wissowa and W. Kroll, XI. 2. (cols. 1338-40). Stuttgart: 1922 [RE].
Robert, Louis: Etudes Anatoliennes. Paris: 1937.
Robert, Louis: Opera Minora Selecta, 4 vols. Amsterdam: 1969-74.
Russell, A.: The Natural History of Aleppo, 2 vols. London: 1794 (reprint Farnborough: 1969).
Schmidt, Max C. P. "Die Fragmente des Mathematikers Menaechmus". Philologus 42, 72-81 (1884).
Schoy, Karl: Die Gnomonik der Araber. Berlin and Leipzig: 1923 (Die Geschichte der Zeitmessung und der Uhren, Bd. I Lf. F).
Schwyzer, Eduard: Griechische Grammatik, 2 vols. München: 1939, 1950 (Handbuch der Altertumswissenschaften II. 1).
Sezgin, Fuat: Geschichte des Arabischen Schrifttums [GAS], Band V, Mathematik bis ca. 430 H. Leiden: 1974.
SIG³: Sylloge Inscriptionum Graecarum (ed. W. Dittenberger). 3rd edn. Leipzig: 1915-24.
Simplicius: In Aristotelis Physicorum Libros Quattuor Priores Commentaria (ed. H. Diels). Berlin: 1882 (Commentaria in Aristotelem Graeca IX).
Sircar, D. C.: Indian Epigraphy. Delhi-Varanasi-Patna: 1965.
Sprenger, A.: Two works on Arabic Bibliography. Calcutta: 1849 (Bibliotheca Indica VI).
Strabo: Géographie Livre II, Vol. I. 2 (ed. G. Aujac). Paris: Budé 1969.
Suidas: Lexicon (ed. Ada Adler), Pars I. Leipzig: 1928.
Suter, Heinrich: Die Mathematiker und Astronomen der Araber und ihre Werke. Leipzig: 1900. (Abhandlungen zur Geschichte der Mathematischen Wissenschaften X). [abbreviated as "Suter"].
Suter, Heinrich: Über die Ausmessung der Parabel von Thâbit b. Ḳurra al-Ḥarrânî". SB Phys.-med. Soz. Erlangen 47, 65-86 (1915)
Ṯabit b. Qurra: Arabische Übersetzung der 'Ἀριθμητικὴ Εἰσαγωγή des Nikomachos von Gerasa (ed. W. Kutsch). Beirut: n. d. [ca. 1958].
Tannery, Paul: "La Cissoide et Diocles". Mémoires scientifiques II. Toulouse and Paris: 1912, pp. 43-47.

Tannery, Paul: "Eutocius et ses contemporains". Mémoires scientifiques II. Toulouse and Paris: 1912, pp. 118-36.
Themistius: Librorum de Anima Paraphrasis (ed. R. Heinze). Berlin: 1899 (Commentaria in Aristotelem Graeca V iii).
Themistius: An Arabic Translation of Themistius' Commentary on Aristoteles de Anima (ed. M. C. Lyons). [London:] 1973 (Oriental Studies II).
Toomer, G. J.: "Apollonius of Perga". Dictionary of Scientific Biography I. New York: 1970, pp. 179-93.
Toomer, G. J.: "The Mathematician Zenodorus". Greek, Roman and Byzantine Studies 13, 177-92 (1972).
Toomer, G. J.: "The Chord Table of Hipparchus and the Early History of Greek Trigonometry". Centaurus 18, 6-28 (1973).
Toomer, G. J.: "Hipparchus on the Distances of the Sun and Moon". Archive for History of Exact Sciences 14, 126-42 (1975).
Trowbridge, Mary Luella: Philological Studies in Ancient Glass. Urbana: 1930 (University of Illinois Studies in Language and Literature XIII, 3-4).
Voorhoeve, P.: Handlist of Arabic Manuscripts. Leiden: 1957 (Bibliotheca Universitatis Leidensis, Codices Manuscripti, VII).
Wainwright, G. A.: Balabish. London: 1920 (Egypt Exploration Society, Memoirs, 37).
Westerink, L. G.: "Elias on the prior Analytics". Mnemosyne 14, 126-39 (1961).
Wiedemann, Eilhard: Aufsätze zur Arabischen Wissenschaftsgeschichte, 2 vols. Hildesheim: 1970 (Collectanea VI).
Wiedemann, Eilhard: "Ibn al Haiṯams Schrift über die sphärischen Hohlspiegel". Bibliotheca Mathematica 3 F, X, 293-307 (1910).
Wiedemann, Eilhard: "Einleitungen zu arabischen astronomischen Werken". Das Weltall 20, 21-26, 131-34 (1919-20).
Woepcke, F.: L'Algèbre d'Omar Alkhayyâmî. Paris: 1851.
Woepcke, F.: "Analyse et extrait d'un recueil de constructions géométriques par Aboûl Wafâ". Journal Asiatique 5 sér., 5, 309-59 (1855).
Wright, W.: A Grammar of the Arabic Language, 3rd edn., revised by W. Robertson Smith and M. J. de Goeje, 2 vols. Cambridge: 1896, 1898.
Zeuthen, H. G.: Die Lehre von den Kegelschnitten im Altertum. Copenhagen: 1886 (reprint Hildesheim: 1965).

Index of Technical Terms

This index contains all terms relating to mathematics, astronomy and technology which occur in the Arabic text of "On Burning Mirrors", arranged alphabetically by roots. All occurrences are listed, except that for those terms occurring a large number of times only a representative selection of instances is given, followed by "al." (other occurrences), "fr." (occurs frequently), or "passim". References are to sections. If a term occurs two, three or more times in a section, this is indicated by a raised "2", "3", etc. following the relevant section number. If a section number is enclosed in angled brackets, that means that the term appears in that section through my restoration or emendation. If a section number is starred, that means that the commentary on that section discusses the term. The forms of the verb are indicated by the conventional Roman numerals. Nouns are characterized as such by the addition of the article "al-" (except where it would be awkward for some reason, e. g. for nouns in the construct case), whether or not they have the article in the passages referred to.

ا — ب

ا

ب

ث

ج

ح

خ

د

ذ

ر

ز

س

ض

ط

ظ

ع

ف

ق

ق — ل

ك

ل

م

ن

ه

و

GPSR Compliance

The European Union's (EU) General Product Safety Regulation (GPSR) is a set of rules that requires consumer products to be safe and our obligations to ensure this.

If you have any concerns about our products, you can contact us on ProductSafety@springernature.com

In case Publisher is established outside the EU, the EU authorized representative is:

Springer Nature Customer Service Center GmbH
Europaplatz 3
69115 Heidelberg, Germany

Batch number: 10620881

Printed by Printforce, the Netherlands